Freaks of Nature

FREAKS
of
NATURE

AND WHAT THEY TELL US ABOUT
DEVELOPMENT AND EVOLUTION

MARK S. BLUMBERG

OXFORD
UNIVERSITY PRESS

OXFORD
UNIVERSITY PRESS

Great Clarendon Street, Oxford ox2 6dp

Oxford University Press is a department of the University of Oxford.
It furthers the University's objective of excellence in research, scholarship,
and education by publishing worldwide in

Oxford New York

Auckland Cape Town Dar es Salaam Hong Kong Karachi
Kuala Lumpur Madrid Melbourne Mexico City Nairobi
New Delhi Shanghai Taipei Toronto

With offices in

Argentina Austria Brazil Chile Czech Republic France Greece
Guatemala Hungary Italy Japan Poland Portugal Singapore
South Korea Switzerland Thailand Turkey Ukraine Vietnam

Oxford is a registered trade mark of Oxford University Press
in the UK and in certain other countries

Published in the United States
by Oxford University Press Inc., New York

British Library Cataloguing in Publication Data

Data available

Library of Congress Cataloging in Publication Data

Data available

Printed in Great Britain
on acid-free paper by
Clays Ltd., St Ives plc.

ISBN 978–0–19–921305–4

... it is often through the extraordinary that the philosopher gets the most searching glimpses into the heart of the mystery of the ordinary. Truly it has been said, facts are stranger than fiction.

—GEORGE M. GOULD AND WALTER L. PYLE
ANOMALIES AND CURIOSITIES OF MEDICINE (1896)

For my mother, Goldene Zalis Blumberg,
Forever the gal in the red dress

CONTENTS

ILLUSTRATIONS

Freaks of Nature

Introduction

Zoos have changed a lot since I was a child. Gone are the sterile, cramped cages and other overt signs of animals in captivity, trapped, often alone, fed their required daily diets but little nourished in other ways, most of them disconnected from the habitats—be they jungles or forests, mountains or savannahs, rivers or seas, ice floes or swamps—in which their counterparts in the wild are born and live their lives. The old zoos could be sad places. They testified to the human proclivity to possess the exotic, the rare, and the unusual while managing to deplete them of their full curious richness by sticking them under glass or behind bars or in a tiny pond, each an "exhibit" plainly labeled for the viewing public to see: mountain gorilla, Asian elephant, Emperor penguin, python.

Most zoos today are certainly more humane and more respectful of the animals in their charge. But even the best of zoos now, with their naturalistic surroundings and cleverly disguised barriers, are still not ideal settings for observing the richness of animal life in all its forms. They can provide only a rough-and-ready sketch of an animal—its size, shape, and color—and the basics of its movement. Moreover, because each animal we see is meant to accurately represent the species to which it belongs, zoos aim to present us with the most exquisite specimens possible—the archetypal Bear, Lion, and Monkey.

Archetypes engender misperceptions. They feed the illusion that, from the moment of conception, nature had a goal in mind. Breeders of pedigree dogs are prisoners of the archetype, devoting incalculable time and energy to creating the perfect Affenpinscher or Yorkshire terrier. In their world, even the slightest blemish is enough to banish a dog to permanent exile.

The pursuit of perfection, canine or otherwise, inevitably butts heads with reality. The world is messy, and nature is unwieldy, unpredictable, and vastly more imaginative than we can ever truly capture—or ward off—with our blemish-free archetypes. Left to its own devices, nature always takes exception to the rule, undermines the archetype, and reminds us that our ideas about what is natural and what we should do to correct nature's "imperfections" are as sound as a sandcastle battered by a rising tide. And nothing batters those ideas with more gusto, more shock and awe, than the creature in nature that is malformed or otherwise anomalous: the freak.

Unfortunate term, *freak*, but that and *monster* are the very colloquialisms that scientists and philosophers have used to describe the odd and the grotesque throughout history. These and similar terms have become common parlance as well, saying more about our overwrought fascination with the unexplained and about our skewed notions of perfection than about oddly formed individuals themselves. When applied to odd-looking human beings, these words also say a lot about our desire to draw a sharp line between them and us. They are freaks. We, however, are just ordinary folks who popped out of the womb perfectly formed, all fingers and toes and everything else accounted for and in the right place, shocking no one because, after all, that's how our bodies are *supposed* to look.

I recently introduced some friends, via computer, to Abigail and Brittany Hensel, the world's most famous conjoined twins. The Hensels are the rarest of the rare—conjoined twins with two heads but just one body. Watching these girls swim, ride a bike, and dribble a basketball, my friends were as stunned as I was when I first saw them in action. One friend asked: *How do they do that?*

In a world gone wobbly for archetypes, we cannot imagine how these twins do that. We cannot imagine how their body and behavior jibe. We do not have a clue how they have managed to function so well and even thrive, conjoined, since birth—or rather, since *before* birth. *Amazing. We* could never live that way. *We* could never do what they have to do to get by every day.

On the other hand, what *we* do is no less amazing after we pop out of the womb perfectly formed, all our parts where they should be, our single heads attached to our single bodies. We function. *How?*

Our mystification at the Hensel twins and their ability to harmonize their uniquely assembled body parts is informative: It exposes how little we understand about physical growth and development, whether in individuals like Abigail and Brittany or in individuals who are normally constructed. It also exposes an unbalanced view that separates the anomalous from the normal, as though one descended inexplicably from outer space and the other was born fully developed according to prescribed earthbound rules. Obviously, the Hensel twins are as earth-bound as the rest of us, and that fact alone should tell us something about the nature of nature. Nary an archetype in the picture, but surely a diversity of unique lives.

It is the aim of this book to explore that diversity by focusing on some of its oddest representatives. It is also to conjoin the worlds of the anomalous and the normal—to show how the Hensel twins and others like them in the animal kingdom reveal the remarkable flexibility inherent in all individual development, the intimate connection between development and evolution, and the nature of development itself. These "freaks," through both their bodies and their behavior, point to underlying processes at work in nature that are at once complex, wondrous, often peculiar, sometimes comical, and always perfectly natural. They have much to tell us about ourselves. For in the larger, unfolding scheme of things, we are all extraordinary, all strange—freaks every last one of us. Some of us just happen to be more notable, with a particularly interesting story to tell.

———

As WE LOOK at developmental anomalies like the Hensel twins, few of us would think to ask: *Who designed them that way?* Certainly no proponents of "intelligent design" would ask such a question, especially if they glanced at the conjoined twins—human and minnow—depicted below. As we will see, there are many such anomalies that betray shared, ancient mechanisms, their appearance alone challenging the notion of a competent and benevolent Designer.

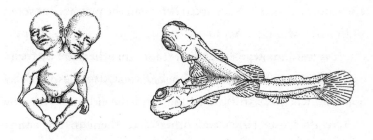

Two-headed conjoined twins look similar in humans (left)
and minnows (right).

For their part, evolutionists argue that the natural world only *appears* well designed. They correctly point to the many instances where the constraints of evolutionary history have necessitated imperfect, even absurd designs that are nonetheless quite functional. For this reason, evolutionists appreciate errors. They understand, for example, that our eye's blind spot reflects a wiring flaw that a competent designer would have avoided. Thus, we say that our eyes betray a process of *apparent* design—a process shaped by natural selection, constrained by history, and far from flawless.

Blind spots and developmental anomalies present an instructive contrast. Each signifies error and imperfection. However, whereas the blind spot has become the darling of contemporary evolutionists taking aim at "intelligent design," these same evolutionists have turned a blind eye to anomalies. For them, developmental anomalies are dysfunctional throwaways, incompatible even with the notion of *apparent* design.

To capture the significance of blind spots and other design flaws, evolutionists have replaced the Designer with the *tinkerer*— "a tinkerer who does not know exactly what he is going to produce but uses whatever he finds around him whether it be pieces of string, fragments of wood, or old cardboards"[1]—a useful metaphorical shift to be sure. As traditionally rendered,[2] the tinkerer only makes incremental adjustments to slightly variable forms and decides which ones to keep.

Not coincidentally, the metaphorical tinkerer fits neatly with the perspective favored by Charles Darwin and his closest adherents. According to them, small, random variations provide the material on which natural selection acts. The result is a continuum of life forms that can be gently nudged in this or that direction to produce the seemingly separate species that we all recognize.[3] But this incrementalist perspective, and the tinkerer that personifies it today, does not quite capture the biological significance of conjoined twins and other developmental anomalies.

Unlike the Darwinians, William Bateson saw the significance of anomalies. Writing at the end of the nineteenth century, he argued that the Darwinian notion that selection acts only on incremental variants was deeply flawed. "Species," Bateson asserted, "are discontinuous; may not the Variation by which Species are produced be discontinuous too?"[4] Accordingly, Bateson catalogued hundreds of extreme variants—monstrosities—with the aim of elevating their status within evolutionary theory. He believed, although at the time he could not prove, that mechanisms internal to the developing embryo—mechanisms inherent in the biological material—critically contribute to the variety on display in nature.

Despite Bateson's best efforts, monstrosities—and the developmental mechanisms presumed to give rise to them—were largely ignored over the ensuing decades. Anomalies would be dismissed as evolutionary dead ends—unfit and irrelevant. Moreover, with the rise of Mendelian genetics at the beginning of the twentieth century, the processes of inheritance that, by means of genes, transmit characters or traits across generations would come to be seen as qualitatively different from those involved in the development of those same characters.[5] Because of this "rupture" between genetic inheritance and developmental mechanisms—between genotype, or the presumed internal genetic constitution of an organism, and phenotype, the observable characteristics of that organism; between nature and nurture—mainstream evolutionary biologists throughout most of the twentieth century would come to view embryonic development as a process to be "largely dismissed as interesting but not informative for evolutionary change...."[6]

Even as these attitudes dominated the field, some biologists still believed that conceiving of evolutionary change without factoring in the developmental contributions to it produced an incomplete picture.[7] But it took time for this developmental perspective to reassert itself. Today, Bateson's views are being dusted off and reexamined,[8] and the rupture between inheritance and development is finally being mended. Indeed, in *Developmental Plasticity and Evolution*, Mary Jane West-Eberhard seeks nothing less than a new evolutionary synthesis of inheritance and development—one "that gives attention to both the transmission and the expression of genes."[9] Her perspective is informed by the insights of Darwin, of course, but also of Bateson, who would have appreciated her section entitled *In Praise of Anomalies*. In that section, West-Eberhard writes:

> To see pattern and extremes in variation is to glimpse the world as seen by natural selection. It is not a world of uniformly tiny, mutationally based, or exclusively quantitative variants. Rather, it is one full of recurrent developmental anomalies

that vary in accord with the genetic makeup of individuals and also with their environmental circumstances.[10]

She continues:

> Unusual variation is abnormal, at least in the sense of being rare, and sometimes even grotesque. But anomalies represent new options for evolution....[11]

The "options" that anomalies offer are embedded within the mechanisms of development. Anomalies help us to glimpse these mechanisms with a clarity that can be lost by focusing on well-formed organisms alone. Consider again the picture of the two-headed humans and minnows. That such strikingly similar forms can arise in these distantly related species testifies to the action of ancient developmental mechanisms. Moreover, as we will see, even extreme anomalies such as these can be produced with surprising ease through the subtlest kinds of developmental adjustments. Thus, it is not difficult to see how diverse forms of life on our planet—including the novel and the anomalous—could have emerged through the evolutionary "tweaking" of developmental mechanisms.[12]

The boundary between novelty and anomaly is a fuzzy one. After all, bodily forms considered anomalous in one species can signify novelty in another. For example, some humans are born without arms and legs, but all snakes are born that way. We may properly characterize limblessness as an anomalous feature in an individual human (in relation to other humans), but as a novel feature of snakes (in relation to other reptiles). But such characterizations do not move us any closer to an appreciation of the shared developmental processes that underlie limblessness in humans and snakes.

This book explores the biological significance of developmental anomalies of various sorts and, in so doing, sheds light on "the evolutionary implications of development and the developmental implications of evolution."[13] Toward that end, we will first consider the historical significance of "freaks" and "monsters" and their role in the historically rocky relationship between developmental and evolutionary perspectives. It is in the context of that relationship that we will return to Darwin and Bateson to discuss their views in greater detail; we will also meet other key figures before and after them whose work has contributed to the literature on this subject. Then we will look at the core principles underlying development itself, in terms of bodily form and function and in terms of the developmental mechanisms that give rise to them. Within that broad framework, we will also look at how anomalous forms can emerge from even subtle changes in the timing of embryonic developmental processes and how both anomalous individuals—whether conjoined, limbless, ambiguously gendered, or otherwise oddly formed—and entire freakish species move about, interact with, and survive in the world.

This is not a textbook, scholarly treatise, or historical survey, although it does draw extensively on both science and history to describe what is known and not known about anomalous beings.

I do not claim to present here a grand new theory. Rather, I aim to use developmental anomalies—in all their forms—as leverage to help the reader gain perspective on a subject that even experts find challenging. However, it is important to emphasize that anomalies represent only one path through the complex terrain that is development and evolution. Thus, my choice to focus here on anomalies is in part pedagogical, and in part principled: Anomalies are difficult to ignore; they matter for development and they matter for evolution; they reveal hidden capabilities and processes in individuals and groups, and in bodies and behaviors; and they are indispensable weapons in the battle to break the spell of "designer thinking."[14] Thus, despite their characterization as errors of nature, the anomalous—when properly considered—force us to confront and correct those errors in our thinking that often impede scientific insight and progress.

Our culture has its archetypes of beauty, intelligence, and athleticism—those ideal standards toward which many of us strive. Archetypes, then, are creatures of our imagination that cloud our conceptions of nature and waft like some hot-air balloon above the fray of the real world. But like ballast, those creatures that we see as odd, abnormal, defective, and freakish give meaningful weight to the balloon and help us stay close to solid ground. When we jettison them, we set ourselves hopelessly adrift on a journey of illusion.

ONE

A Parliament of Monsters

On the Breadth and Scope of
Developmental Anomalies

Treasure your exceptions!.... Keep them always
uncovered and in sight. Exceptions are like the
rough brickwork of a growing building which
tells that there is more to come and shows
where the next construction is to be.
WILLIAM BATESON (1908)[1]

In the spring of 1940 in the Dutch city of Utrecht, shortly
before Hitler imposed his vision of Aryan perfection on
Holland, a horribly malformed goat died. Much has been written about this goat,[2] whose accidental death at just one year of
age cut short an extraordinary life. Because although this animal
had been born without forelegs, it had, despite its deformity,
developed the ability to walk—upright. In Latin, such a creature is called a *lusus natura*—a joke of nature. A freak, a monster.
A violation of God's intentions, of the natural order of things,
of the evolutionary forces that shaped and refined its ancestors.
Clearly, this was one special goat.

One also hears of humans overcoming birth defects and
other disabilities in remarkable ways. Christy Brown, the

Irish artist and writer whose exceptional ability to cope with cerebral palsy was dramatized in the movie *My Left Foot,* provides compelling testimony to individual adaptability in the face of severe challenge. Perhaps even more compelling are wheelchair-bound athletes, such as those depicted in the documentary film *Murderball,* whose competitive drive lifts them beyond their physical limitations.

Such tales of triumph over physical challenge are among our most treasured sources of inspiration. Much less treasured, however, are those anomalous productions of nature that challenge our capacity for tolerance and understanding and that all too often horrify and disgust us. Still, we find these oddities undeniably fascinating. Why? Is it that our eyes are naturally drawn to novelty, toward those objects that violate our expectations of how things *ought* to look? Also, why is our initial response so often one of fear and, all too often, cruelty? Perhaps it was fascination tinged with fear that once made acceptable the corralling of freaks within the controlled environments of fairs and circuses. P. T. Barnum grasped this dynamic. He was a genius at extracting money from a Victorian public that was "mad about monstrosity."[3] Indeed, Barnum's "specimens," collected and displayed in his time, remain familiar to us by the names that he bestowed upon them: Tom Thumb, the Elephant Man, the Bearded Lady, the Siamese Twins.

Barnum's crassness is countered by the sensitivity of Diane Arbus, the great American photographer who memorably captured so many examples of human eccentricity. She once commented that you "see someone on the street and essentially what you notice about them is the flaw."[4] Starting from this perceptual bias, she focused on marginalized individuals, "[n]ot to ignore them, not to lump them together, but to watch them, to take notice, to pay attention."[5] Her haunting photographs—of giants and dwarves, twins and triplets, transvestites, nudists, and sadomasochists—permit the kind of reflection that the frenzy of a freak show cannot. As we gaze upon her subjects, we glimpse their individuality, not just the group to which they belong.

In the menagerie that Barnum paraded before a mesmerized public and in the individuality that Arbus captured on film, we grasp the breadth and scope of freakishness. But as we will see in this book, there is much more to learn about these marvels than either Barnum or Arbus realized. Beyond voyeurism and fine art, freaks provide ready access to some essential truths about the development and evolution of animal form and behavior, and about the individual potential within each of us.

The lives at the heart of our story are tethered precariously to our planet. Like a polar bear stranded on a floating ice sheet, Earth sits alone in our solar system, a sign of life in a largely frigid wasteland. The emergence of life, then, is a rare event. But even the discovery of life beyond the perimeter of our planet would not alter the fact that, in a universe comprising more dead space than living matter, every creature in our midst is a freak. We are, moreover, freaks with a history—a history whose origins can be traced to those ancestral moments when a few simple molecules marched across the gray zone that separates animate from inanimate matter.

The hallmark of terrestrial life—and perhaps the hallmark of life wherever we might find it—is its diversity. There exist millions of species of bacteria, hundreds of thousands of species of plants, one million species of insects, and thousands of species of fish, amphibians, reptiles, birds, and mammals. But diversity does not stop at the boundaries that separate a human from a chimpanzee, a lion from a tiger, or a canary from a cowbird. Within each species, we can find remarkable diversity as well. To appreciate this, we need only consider the uniqueness of each human face or the distinctive color patterns of individual zebras. But in addition to these "natural" variations among members of a species, other kinds of variations are so extreme that they challenge our ability to comprehend their place within the natural order of things.

So we can imagine the confusion that must have gripped our ancestors when, in a world where every birth would seem a miracle, they confronted an infant born with a single torso supporting two perfectly formed heads. Such seminal events in a small community would naturally be suffused with special meaning, and were often interpreted as the delivery of a message from the gods. This ancient curiosity about the meaning of monstrous births is documented in a 4000-year-old cuneiform clay tablet, discovered near the Tigris River. The tablet lists sixty-two human malformations, each assigned an omen, such as death, destruction, or starvation. Little wonder, then, that the word *monster* likely derives from the Latin word *monere*, meaning *to warn*.

However, malformed infants were not always perceived as mere conduits of communication from a godly realm. Occasionally, they became, if not gods themselves, then the inspiration for godly images.[6] For example, Janus, the Roman god of gates and doors—of beginnings and endings—comprises two male heads, fused at the back, each face gazing metaphorically into the past and the future. Janus's seemingly mythical anatomy eerily mirrors those real-world conjoined twins that have fused heads with oppositely oriented faces.

Even Atlas, that imposing figure of Greek mythology, forced by Zeus to carry the world on his shoulders, may have been inspired by the appearance of an infant with a particular malformation of early development.[7] In this disorder, an abnormal gap in the back of the skull enables the growth, called an *encephalocele,* of the underlying membranes and brain tissue, resulting in a bulbous mass of tissue resembling a globe perched on the shoulders. Similar embryological parallels may underlie other mythological creatures, such as mermaids and cyclops.

But one will search in vain for a simple and straightforward narrative detailing the social reaction to monstrous births throughout human history. Children with congenital malformations may have inspired the sacred imagery for Atlas and Janus, but they also stirred a secular disgust that often resulted in their death—in ancient Rome they were drowned in the Tiber River.[8] More recently, the life of an eight-limbed Indian

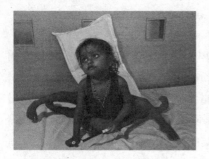

Lakshmi Tatma (left) in 2007 before the surgery to remove her extra limbs.
The multi-limbed Hindu god, Vishnu (right).

girl may have been spared by portentous timing: she was born on the very day set aside to celebrate Vishnu, the multi-limbed Hindu god. Such has been the discordant response of our species to the monstrous: fascination and fear, admiration and contempt, consecration and desecration.

Because there have always been too many sinners and not enough monsters, monstrous births were never easily pitched to the people as punishment for individual moral failings. Moreover, the fact that domestic animals also produce monstrous births made it more difficult for religious leaders to point the finger of blame at human parents of monstrous infants. Rather, more typically and more effectively, these tragic events were cast in a symbolic light—as signs that God was paying attention to our activities and was not at all pleased with what He saw. Thus, by encouraging us to reflect on our behavior, monsters were meant to move us away from sin. Their use as symbols emerged precisely because communities were small and each monstrous birth was so unusual. But as communities grew, as in Europe, monstrous births were increasingly frequent, thereby diluting them of their uniqueness and significance.

The advent of printing had a particularly dramatic impact on changing attitudes toward monsters. For example, in Elizabethan England during the sixteenth and seventeenth centuries, birth announcements of the monstrous variety were often printed in verse form on single sheets of paper—called *broadsides*—and advertised to villagers and town folk through song.[9] The broadsides conveyed accurate information about each birth, often including detailed family information and illustrations of the deformed children. For the children and their parents, celebrity status soon followed and so did compensation from the hundreds of paying visitors encouraged by the broadsides to see these families for themselves. Even if some viewed the infants as proper punishment for the moral decrepitude of their parents, that did not stop them from paying the price of admission.[10]

When parents began taking their "show" on the road, they were able to reach a broader audience. This market for monsters would expand during the eighteenth and nineteenth centuries, when the European and American publics' passion for monsters reached its zenith. William Wordsworth, the English romantic poet, captured this passion in his autobiographical collection, *The Prelude*. Writing in the early 1800s, Wordsworth described the Bartholomew Fair, held near London each year at the end of August:

> All moveables of wonder, from all parts,
> Are here—Albinos, painted Indians, Dwarfs,
> The Horse of knowledge, and the learned Pig,
> The Stone-eater, the man that swallows fire,
> Giants, Ventriloquists, the Invisible Girl,

The Bust that speaks and moves its goggling eyes,
The Wax-work, Clock-work, all the marvellous craft
Of modern Merlins, Wild Beasts, Puppet-shows,
All out-o'-the-way, far-fetched, perverted things,
All freaks of nature, all Promethean thoughts
Of man, his dulness, madness, and their feats
All jumbled up together, to compose
A Parliament of Monsters.

Today, the thought of such spectacles and the crowds that flocked to them appall us. We have lost our taste for the exploitation of the unusual. Or so we like to tell ourselves.

HITCHES UNDER HEAVEN

In a culture where God was credited for *all* His creations, monsters were to be counted among them. But as monsters became more familiar to common folk and scholars during the sixteenth century and beyond, it became increasingly difficult to see in them the handiwork of God. No longer isolated and rare, monsters demanded explanation. A science of monstrosity was on the horizon.

In 1573, a French surgeon named Ambroise Paré published *On Monsters and Marvels*. This book, considered by many historians to be a watershed in the scientific discussion of monsters, was notable for its author's focus on the *causes* of monstrous births. However, as the opening passage of Paré's book makes clear, he could not entirely shed his links with the prescientific past:

> There are several things that cause monsters.
> The first is the glory of God.
> The second, his wrath.
> The third, too great a quantity of seed.
> The fourth, too little a quantity.
> The fifth, the imagination.
> The sixth, the narrowness or smallness of the womb.
> The seventh, the indecent posture of the mother, as when, being pregnant, she has sat too long with her legs crossed, or pressed against her womb.
> The eighth, through a fall, or blows struck against the womb of the mother, being of child.
> The ninth, through hereditary or accidental illnesses.
> The tenth, through rotten or corrupt seed.
> The eleventh, through mixture or mingling of seed.
> The twelfth, through the artifice of wicked spital beggars.
> The thirteenth, through Demons and Devils.[11]

Note the curious mix of the material and the spiritual, of high-minded moral judgment and mundane physical cause. In addition, his focus on early development—from fertilization through birth—is striking. But this focus could not compensate for his ignorance about development. Thus, Paré may have been interested in understanding the causes of monstrous births—indeed, he organized his book around that theme—but he was not equipped to do so.

Paré's distinction between monsters and marvels clearly places him in an earlier age. As he described them, *monsters* were "things that appear *outside* the course of Nature..., such as a child who is born with one arm, another who will have two heads."[12] In contrast, *marvels* were "things which happen that are completely *against* Nature as when a woman will give birth to a serpent, or to a dog." In other words, in a world typified by the *natural*, monsters were *unnatural* and marvels *supernatural*.[13]

Hampered by a limited understanding of biology, Paré was trying to make sense of the disparate observations and illustrations of others, many of them inaccurate and unreliable. Not surprisingly, then, no obvious biological basis exists for Paré's distinction between monsters and marvels. Certainly, Paré's scheme is not reducible to a simple division between the biologically plausible and implausible. Such a division makes no sense. For example, the birth of an otherwise normal dog to a human female is just as biologically implausible as the birth of a human-dog hybrid, and yet Paré distinguished between the two, classifying the former as a marvel and the latter as a monster. He even recounts as fact the birth of a centaur-like monster, supposedly born in 1493, whose upper torso resembled, "from the navel up," its human mother and whose lower parts derived from its canine father.[14]

Drawings of a child purported to have been
born in 1493 to a human mother and canine father (left)
and a man "from whose belly another man issued" (right).

As we will see, the individuals who were to lay the groundwork for a modern science of developmental malformation
would dispense with Paré's distinction between monsters and
marvels. As for Paré, we should forgive his naiveté. After all,
many of the creatures illustrated in his book seem just as fantastic as centaurs but, as we now know, actually exist: Humans
with two heads and a plethora of limbs *exist*, as do "two twin
girls joined together by their foreheads,"[15] a man "from whose
belly another man issued,"[16] and individuals with a normally
sized head and torso but with grotesquely foreshortened arms
and legs. Without a solid scientific foundation upon which to
sift evidence, Paré's judgment was a bit shaky.

Despite his shortcomings, Paré bridges our ancient and modern fascination with monsters. Soon after his work, marvels would disappear from the discussion and monsters would come to be viewed as mistakes, errors, slips, and defects. Writing in 1616, forty-three years after Paré, the Italian scholar Fortunio Liceti clearly enunciated this break with the past. Unlike Paré, Liceti had no use for marvels and, moreover, he moved confidently to expunge any hint of the supernatural from the world of monsters:

> A monster is a being *under heaven* that provokes in the observer horror and astonishment by the incorrect form of its members, and is produced rarely, begotten, by virtue of a secondary plan of nature, as a result of some *hitch* in the causes of its origin.[17]

Going even further, he added: "It is unbelievable that God produces monsters in order to warn men of imminent catastrophes."[18]

Uncovering the "hitch" that might move a developing animal toward "a secondary plan of nature" required a science that was not yet available to Liceti. Two hundred years later, in the late eighteenth century, Etienne Geoffrey Saint-Hilaire, the French anatomist and natural historian, founded such a science, and it was his scientist son, Isidore Geoffrey Saint-Hilaire, who gave this field its name: *teratology*, meaning literally the study of monsters (although today the term refers to the investigation of

congenital malformations, particularly those caused by environmental insults). Etienne's initial goal was to classify monstrous
forms in the same way that, earlier in the century, Carl Linnaeus
had classified nonmonstrous forms. As Isidore described his
father's perspective:

> Monsters are not sports of nature; their organization is subject
> to rules, to rigorously determined laws, and these rules, these
> laws, are identical with those that regulate the animal series;
> in a word, monsters are also normal beings; or rather, there
> are no monsters, and nature is one whole.[19]

As Liceti expunged the supernatural from the world of monsters, Etienne expunged the unnatural. Paré's unholy trinity—
the natural, unnatural, and supernatural—was gone forever.

Etienne never attained his more ambitious goal of integrating his knowledge of monsters with a broad understanding of
all animal forms and evolution. Nonetheless, his work propelled the examination of monsters far beyond the fantasy and
prejudice that marked the writings of his predecessors.

As a consequence of Etienne's efforts, by the end of the
nineteenth century when George Gould and Walter Pyle
published their comprehensive examination of teratologi-
cal phenomena, *Anomalies and Curiosities of Medicine*, the science
of teratology had come into its own. Their photographs and
precise illustrations can be shocking to a modern reader in a
way that Paré's cartoonish images cannot. Thus, whereas Paré

Photograph of Laloo and his parasitic twin, a real-world
example of a man "from whose belly another man issued."

described a man "from whose belly another man issued" with
an accompanying sketch, Gould and Pyle offered a vivid photograph of a normal-sized young man named Laloo, dressed in
the costume of a circus performer. Dangling in front of Laloo
is the torso of his child-sized twin, similarly dressed, his head
seemingly stuck inside Laloo's belly. Like a puppeteer, Laloo
holds his brother at the wrists.[20]

Such "parasitic monsters" are so horrifying, so *marvelous*,
that they do not seem real; but they are. So are the many other
bizarre images that fill Gould and Pyle's book: an infant with a

second partial head, a child with three heads, men with double penises, individuals with both male and female genitalia, women with beards, children without limbs, "sirens" (children whose lower extremities are completely fused, giving a mermaid-like appearance), and individuals with double hands, hands with extra fingers, feet with extra toes, or an outgrowth from the base of the spine that resembles a stumpy tail. There is even a woodcut depicting a seated mother nursing an infant at her breast as a standing child sucks on an auxiliary nipple located on her thigh.[21]

What do we possibly gain from such collections of unsettling images, published for all to see and, perhaps, even ridicule? We gain perspective. We see that nature is imperfect, and with this simple insight we illuminate our past, our present, and our future. Indeed, Gould and Pyle saw the value of their volume in exactly those terms. By examining monstrosities, they wrote,

> we seem to catch forbidden sight of the secret work-room of Nature, and drag out into the light the evidences of her clumsiness, and proofs of her lapses of skill,—evidences and proofs, moreover, that tell us much of the methods and means used by the vital artisan of Life,—the loom, and even the silent weaver at work upon the mysterious garment of corporeality.[22]

Around the same time that Gould and Pyle wrote those words, a young biologist named William Bateson was likewise promoting monstrosities. But Bateson went further than

Gould and Pyle and positioned monstrosities within an evolutionary scheme. His efforts, however, would alienate him and his perspective from mainstream biology for many decades.

DEVELOPMENT EVOLVING

It is noteworthy that in *The Origin of Species*, published in 1859, Charles Darwin looked to the embryo for evidence to support his theory of evolution.[23] So too did his zealous and somewhat subversive supporter, Thomas Henry Huxley. A man who welcomed intellectual combat, Huxley published his *Evidence as to Man's Place in Nature* four years after Darwin published *The Origin*. In this slim book Huxley addressed head-on the issue of *human* evolution that Darwin dodged until 1871, with *The Descent of Man.* "The question of questions for mankind," Huxley wrote, "the problem which underlies all others, and is more deeply interesting than any other—is the ascertainment of the place which Man occupies in nature and of his relations to the universe of things."[24] Just three pages after writing those words, Huxley begins his discussion of embryonic development.

Huxley, like Darwin and many others,[25] noted that the embryos of closely related species appear more similar than do the embryos of distantly related species, and that differences between embryos become increasingly apparent as development proceeds. In Huxley's words, the more closely any animals resemble one another in adult structure, the longer and the more intimately do their embryos resemble one another: so that, for example, the embryos...of a Dog and of a Cat remain like one another for a far longer period than do those of a Dog and a Bird; or of a Dog and an Opossum; or even than those of a Dog and a Monkey.[26]

If embryos are nearly identical at the earliest stages of development and only diverge at later stages, then it would follow that evolution proceeds by accumulating new features only at the end of embryonic development. This process, akin to adding new cars to the end of an ever-elongating train, is called *terminal addition*. Twentieth-century embryologists came to recognize that terminal addition is not the sole developmental mechanism underlying evolutionary change.[27] For example, Walter Garstang saw that a "house is not a cottage with an extra storey on the top. A house represents a higher grade in the evolution of a residence, but the whole building is altered—foundations, timbers, and roof—even if the bricks are the same."[28] Today, more detailed embryological observations are buttressing the notion that the earliest embryonic forms can foreshadow novel, species-typical characteristics in adults.[29] Thus, modifications to embryos at *any* stage of development, not just later ones, can effect evolutionary changes in adult characteristics.[30] But for Darwin, terminal addition was *the* essential mechanism of adaptation and evolutionary change.[31]

Moreover, in *The Origin*, Darwin discusses how the breeders of dogs, horses, and other domesticated animals—breeders whom he often consulted for insight and inspiration[32]—mate their animals based solely on characteristics or traits that are evident in adults. Specifically, Darwin emphasizes how these breeders, when judging their animals, are "indifferent whether the desired qualities are acquired earlier or later in life, if the full-grown animal possesses them."[33] In other words, for these breeders it is the product that matters, not the process. It is the destination, not the journey.

But what if the journey does matter? What if the developing embryo is more than a mere intermediary in the transmission of adult forms from one generation to the next? The approach adopted by Darwin's breeders was a bit like looking at a list of World Series champions for insight into the game of baseball. Each entry on the list records that year's champion, but the details that connect champions from year to year—from spring training to opening day through the regular season and into the playoffs—are missing. It is history without the drama.

For anyone interested in the drama, this approach alone would not do. One such person was William Bateson, a pioneering geneticist who was convinced that the embryo has much to teach us about an animal's history, construction, and evolutionary potential.[34]

Born in 1861, Bateson was in his mid-thirties when he entered into a dispute with Darwin's successors concerning continuity and discontinuity among species. The Darwinians, like Darwin himself, considered the concept of a species a mere convenience for characterizing the natural world. Accordingly, although species appear as discontinuous entities, the Darwinians assumed a hidden continuity—just as a single iceberg can have separate peaks piercing the ocean surface. In contrast, Bateson believed that species are as they appear to be Separate. Discontinuous. For Bateson, horses and humans, caterpillars and crickets, salmon and swordfish, and eagles and egrets are truly, not just seemingly, separate. Thus, the notion of a species was far from a mere convenience for Bateson, but rather a foundation upon which a science of biological form and development could be built.

The Darwinian conception of evolution, then, begins with a group or population of animals in which each individual is a variation on a theme. Importantly, the observed variation is incremental and randomly generated, produced without any constraints imposed from within the animal. Thus, the driving force of evolutionary change—natural selection—has to be an *external* one.

But Bateson sensed that not all variation is equally possible—that developmental factors *internal* to the organism *constrain* or *bias* the range of forms that nature produces.[35] Rather than view natural selection as the engine of evolution, Bateson placed his bet on variation. Indeed, in his *Materials for the Study of Variation, Treated with Especial Regard to Discontinuity in the Origin of Species,* published in 1894, Bateson called variation "the essential

phenomenon of Evolution."[36] For him, variation and discontinuity go hand in hand:

> The first question which the Study of Variation may be
> expected to answer, relates to the origin of that Discontinuity of
> which Species is the objective expression. Such Discontinuity
> is not in the environment; may it not, then, be *in the living thing
> itself?* [37]

For Bateson, odd, irregular, and atypical animals reveal the
discontinuity that he wished to expose. Such monsters represent
biological forms uncontaminated by the demands of natural
selection, providing unfettered access to the internal rules and
mechanisms of development. Moreover, as Bateson documented
in *Materials*, monsters were hardly the rare occurrences that many
imagined.

Bateson's nearly 600-page book overflows with oddities:
a fly's antenna that ends in a foot; a man with an extra set
of nipples; a human hand with a double thumb. The vast
majority of his observations highlight subtle, but discontinuous variations in the expression of an animal's basic body
plan. Bateson recorded variations in the segments of earth-
worms, the vertebrae of snakes, and the ribs of humans. He
examined teeth, fingers, and toes; wings and scales. Indeed, in
his quest for the rules of growth and variation, he examined
any anatomical feature that extends, segments, branches, or
divides.

The discontinuous character of monsters, Bateson believed, violated the smooth continuity that Darwin had embraced (but Huxley had not[38]). Darwin himself explicitly considered and then rejected any role for monsters in his evolutionary scheme.[39] For him, they were "mere monstrosities"—abrupt accidents that might emerge under the watchful gaze of breeders, but not adaptive variations produced and propagated and then sustained in response to conditions in the natural environment.[40]

Continuity became a core principle of twentieth-century Darwinism. So much so that when the geneticist Richard Goldschmidt, echoing Etienne Geoffrey Saint-Hilaire before him, asserted in 1940 that spanning the "bridgeless gap" that separates species requires "a completely new anatomical construction" and that this construction can arise "in one step"— he infamously dubbed such creatures "hopeful monsters"—he was vilified.[41] The Darwinians considered monsters, whatever their form and origin, the embodiment of discontinuity, and so they were not to be tolerated. Monsters were to have no privileged place in a continuous world.[42]

Nonetheless, Bateson's observations revealed discontinuities in the developmental assembly of body parts that contradicted the Darwinian expectation of continuous variation. Then, upon the rediscovery in 1900 of Gregor Mendel's experiments, originally conducted some thirty years earlier, on the genetic transmission of characteristics across generations, Bateson hastily recruited Mendel's findings to support his notion of discontinuous evolutionary change. With that, the seeds were sown for a bitter dispute[43] between Bateson, considered by many a stubborn man who enjoyed controversy, and a group of mathematically oriented Darwinians, who themselves were hardly paragons of flexibility. (One of them described evolution as a problem with solutions that "are in the first place statistical, and in the second place statistical, and only in the third place biological.")[44] The ensuing clashes of ideology and personality would not subside until it was demonstrated that Mendelism was not, as Bateson and others had supposed, incompatible with Darwinism. That compatibility gained further traction when in 1930, four years after Bateson's death, the statistician and geneticist R. A. Fisher published *The Genetical Theory of Natural Selection*, merging Mendelian genetics with Darwinian natural selection. With that, the Modern Synthesis—with its emphasis on evolution as a process of slow, incremental changes in the genetic constitution of continuously varying species—was born.

There was much irony in this turn of events. Bateson would have been dismayed to witness the incorporation of his field, genetics, into a worldview that he so fervently opposed. But just as troubling was the omission of development from the Modern Synthesis. This omission rested in part upon the "rupture" that Mendelism created between genetic inheritance and developmental mechanism.[45] Also, the commitment of the founders of the Modern Synthesis to continuity and random genetic variation convinced them that development could be safely ignored.

Goldschmidt and others criticized the Modern Synthesis for its inattention to development,[46] but such criticisms were largely disregarded. After all, the Modern Synthesis "was not so much wrong as it was incomplete."[47]

Then, toward the end of the twentieth century, Bateson's work suddenly became relevant again. The anomalies that he had so carefully described in *Materials for the Study of Variation* were being produced in the laboratory by manipulating a small class of genetic components. Moreover, these components appeared to help regulate the development of animal appendages and body segments across a surprisingly diverse array of species— indeed, in every species studied thus far, including crustaceans, centipedes, mice, and humans.

Without knowing it, Bateson had laid the foundation for a revolutionary rethinking of the molecular basis of evolutionary change. With this new thinking came a new field—christened evolutionary developmental biology, or Evo Devo.[48] After decades of exile, development and discontinuity were welcomed back into the evolutionary fold.

MORE THAN MUTANTS

Despite its many successes, Evo Devo continues an unfortunate tradition of scrutinizing development primarily through a genetic lens. But if development is, as Evo Devo enthusiast Sean Carroll states, "the process that transforms an egg into a growing embryo and eventually an adult form,"[49] then no theory of development can afford to neglect any factor that influences that process. Unfortunately, one of the legacies of a century of gene-focused thinking is a narrowed vision of how development actually happens—how at every step the transformation of an egg into an adult involves local molecular events involving genes, proteins, and other molecules, as well as often-neglected factors such as temperature, gravity, and oxygen.

Consider the seemingly miraculous transformation of a maggot into a fly or a tadpole into a frog. Most of us are familiar with the end points, but not the moment-to-moment processes that link them. These processes are the stuff of development, but are so complex that genes alone are often invoked to simplify and explain them to the exclusion of other, no less real but perhaps less easily manipulated factors that just as critically guide and shape the developmental process.[50] If you radically change the incubation temperature of a duck embryo, for example, you get duck soup.

The overwhelming success of genetic research over the last fifty years has delayed widespread acceptance of a biologically realistic approach to understanding development. Thus, it should come as no surprise that Paré's monsters are today's mutants. Indeed, in *Mutants: On Genetic Variety and the Human Body,* an otherwise elegant, informative, and humane book published in 2003, Armand Marie Leroi focuses primarily on freaks as *genetic* anomalies. Central to Leroi's perspective is the misguided notion that genes are entities with the power to design and control, or that they contain the blueprint from which an animal is built. In fact, on the very first page of *Mutants*, Leroi clearly states this still-popular view: "Our bodies—I hesitate to add our minds—are the products of our genes. At least our genes contain the information, the *instruction manual*, that allows the cells of an embryo to make the various parts of our bodies."[51]

As one developmental biologist has noted,[52] the metaphorical conception of DNA as a program, recipe, or blueprint for development has "an admirable sharpness and punch" that is hard to resist. But if we do not resist it, we can too easily find ourselves on that slippery slope from literary metaphor to supposed fact. Instead, we should recognize DNA for what it is: a molecule that helps provide the raw materials for development. Only with that recognition can we finally embrace a biologically realistic view in which DNA—contrary to the awesome power so often ascribed to it—is seen as one among many contributors to a complex and highly interactive system.

Of course, genetic mutations help explain many human developmental anomalies. But the intense spotlight that we continue to focus upon genes hides a deeper reality about developmental anomalies that, when acknowledged and appreciated, leads us to a richer and more comprehensive understanding of biological form and behavior.

MONSTERS AT THE TABLE

Evo Devo is but one response to the limitations of the Modern Synthesis. The reintegration of developmental and evolutionary biology also occurred through the work of such scientists as the late Stephen Jay Gould, who in 1977 published *Ontogeny and Phylogeny*, an ambitious attempt to explain how development of an individual organism—*ontogeny*—is modified to alter the evolutionary relationships among species—*phylogeny*. Gould's book triggered a resurgence of interest in this problem. For one young scientist, Pere Alberch, the book was particularly life changing.

Charming, outrageous, irreverent—an *enfant terrible* according to one former student[53]—Pere Alberch dedicated his career to synthesizing development and evolution. Born in Spain in 1954 and educated in the United States, he was a professor at Harvard University before he was denied tenure and forced to leave in 1989. He died in Spain from heart failure nine years later at the age of 43.

Of the many wonderful papers that Alberch wrote one stands out for its creativity and fresh perspective. Written in 1989, "The Logic of Monsters" is a gem that sparkles with the passion of a true believer. In this paper, Alberch reevaluated Darwinism's central commitment to continuous variation and the power of natural selection, if given sufficient time, to produce near-infinite variety. In a plea for balance, Alberch highlighted the inherent rules that shape and guide, but also constrain what is possible during the course of development. To bring his argument home, he emphasized the insight to be gained by considering biological forms in their purest sense— even those monstrous forms that plainly exist and that just as plainly fail in the evolutionary struggle for survival.

For Alberch, development is not a river of genetic information flowing inexorably downstream toward the creation of biological form, but rather many rivers, tributaries, and eddies—a turbulent, cyclical process involving gene regulation and protein synthesis. But these complex processes do not produce infinite possibility because buried within them are pattern-generating "rules of development" that "define the realm of possible variation and place limits on the process of adaptation."[54] Alberch believed that monsters are ideal subjects for investigating these inherent rules and patterns. They are ambassadors of pure form and thus deserve a seat at the evolutionary table.

So, like Bateson, Alberch distinguished between the *external* force of natural selection and the *internal* dynamics of development.[55] He contrasted the evolutionary lineage of a species with the developmental history of an individual; the transmission of form across generations with the regenesis of form within each generation; the unconstrained with the constrained; the past with the here-and-now.

But unlike Bateson, who may have tipped the scale too far in the direction of internal factors and extreme variation, Alberch sought balance. He argued that the animal forms that we see around us must result from a combination of internal factors producing a broad range of discrete—not continuous—variations,[56] and external factors selecting among them. But because gene-centered, population-level thinking—enshrined by the Modern Synthesis—invariably escorts us toward the externalist perspective, Alberch appealed for renewed emphasis on the study of form from the internalist perspective. To put this appeal in the starkest terms possible, he thrust monsters forward as

> a good starting point towards such a goal; they represent forms which lack adaptive function while preserving structural order. There is an internal logic to the genesis and transformation of such morphologies [different forms and structures of living organisms] and in that logic we may learn about the constraints on the normal.[57]

Alberch was writing when the pioneering experiments that would establish the field of Evo Devo were fresh and their implications were only beginning to be appreciated. He grasped the significance of identifying a small, finite number of genes that contribute to the construction of similar body parts across a wide array of species. Many Evo Devo enthusiasts see such genetic insights as further evidence of the primacy of genes. In contrast, Alberch saw these new findings as support for his belief in internal constraint, in a relatively small number of cellular combinations and operations available to the developing organism.

Thus, when faced with a novel animal form—for example, a fly with an extra pair of wings—Alberch did not immediately imagine a novel trait produced by some magical mutation. Rather, he saw a *system* that had been modified to produce a novel form. Consequently, he understood the need to determine how *all* of the components of the system—genetic and extra-genetic—interact to produce both typical and novel forms.

Ultimately, appreciating the intimate connection between development and evolution requires a perspective that differs from that associated with traditional Darwinism. Recall Darwin's description of the indifference of animal breeders to "whether the desired qualities are acquired earlier or later in life." But evolution is not indifferent. On the contrary, the notion that selection—whether natural or artificial—acts only to preserve individual anatomical traits is misleading. For example, when a farmer breeds dogs to herd or cows to produce milk, he is selecting for much more than just herding and milk production. In particular, what is being preserved across generations is the *developmental means*[58] to produce the desired outcome, as well as all of the other features that come along for the ride.[59] Scientists and breeders may often ignore such nonobvious and unpredictable side effects of selection, but that does not diminish their significance for an individual and its descendants.

The focus on developmental *process* rather than genetically preordained *product* is the hallmark of the perspective known as *epigenetics.*[60] Epigenetics—and the foundational concept of *epigenesis,* which can be traced back to Aristotle—embodies the interactive nature of development. When Conrad Waddington coined the modern term in the 1940s,[61] he conjured the image of a sloped terrain of hills and valleys, upon which a rolling ball—representing the development of an embryo—courses down one path or another.[62]

Waddington went further with his metaphor to suggest that the landscape itself was like a piece of fabric draped over a complex interconnected system of ropes anchored to pegs. Tension on the ropes produced the valleys in the landscape's fabric above, and, as he imagined must be the case with any genetic system, a change in the tension in one location would have cascading influences on the shape of the entire landscape. Change this *epigenetic landscape* and development will course down a different path.

Though Waddington's visual image would eventually prove a powerful one, it could never compete with the more easily grasped "sound bites" that geneticists were offering. Epigenesis cannot be reduced to a sound bite. Indeed, after so many decades of the steady drumbeat of genetics, epigenesis can sound too nuanced, perhaps even mystical. But it is neither. It is the way things actually are—elaborate and complex, tortuous and convoluted.

The epigenetic perspective is one among several core principles that define a broad conceptual scheme now referred to as *Developmental Systems Theory* (DST).[63] The proponents of DST seek to replace the simplistic single-cause, gene-centered thinking that is so prominent today. They note how genes have been ascribed the dual role of being the designer of development *and* the sole means by which design is inherited by members of the next generation. This double duty foisted onto genes is a double mistake: Having placed so much control into the hands of mere molecules, we must now work hard to distribute that control more equitably.

DST aims for this equitable distribution by emphasizing how genes are only one part of a complex, interacting, developmental dynamic. Moreover, genes are not the sole means of inheritance. On the contrary, the list of *nongenetic mechanisms of inheritance* is now quite long, and it includes inherited features of the external environment—including gravity and oxygen, jungles and oceans, nests and burrows, and even language—as well as a variety of inherited nongenetic molecular processes that influence gene expression and other developmental events. This notion—that the environment consistently and reliably shapes development across generations—can no longer be credibly denied. Still, many remain skeptical, leading one prominent evolutionary thinker to rank this skepticism "among the oddest blind spots of biological thought."[64]

When considered in its entirety, DST offers a balanced and realistic perspective of development that is sorely needed today. It views development as a process in which all components participate in the moment-to-moment construction of the individual, generation after generation, genesis and regenesis.

Consistent with DST, Alberch criticized the notion that information flows from the genes like water from a faucet. Instead, he saw nested cycles of interactions—loops within loops within loops—that regulate the expression of genes both within each individual cell and among all the cells of the body. Thus, gene expression results in the production of proteins, proteins participate in the physico-chemical interactions within cells, and these interactions ultimately result in the production of new tissue. In turn, the physical orientation of tissues modulates the next round of gene expression. And round and round it goes.

HARD AND SOFT PARTS

Freaks sometimes arise through human intervention, both real and imagined. Prehistoric cave drawings depicting human–animal hybrids testify to an ancient fascination with impossible creatures. In 1894, H. G. Wells manipulated this fascination to great effect in his darkly disturbing novel, *The Island of Dr. Moreau*. In his story, Wells introduces us to a physician, working secretly in a laboratory on a secluded island, brutishly attempting to "humanize animals" through surgical intercession—"animals carven and wrought into new shapes."[65] Moreau's attempted demonstration of the "plasticity of living forms"[66] managed only to produce "grotesque caricatures of humanity."[67] Some people feel that our current obsession with plastic surgery for cosmetic purposes—itself emblematic of the human tendency to tinker with what nature provides, in pursuit of beauty or perfection or simply to fill in nature's gaps—produces similarly grotesque results.

If, through Moreau, Wells made an error, it was in not fully comprehending the difficulty of transforming animals once they have reached adulthood. Surgical intervention in adults cannot guarantee the full integration of parts that development accomplishes so effortlessly. The same can be said for other interventions. For example, steroids may enhance athletic performance, but their harmful side effects betray an interdependent system whose balance has been upset. In contrast, as embryologists know well (and as we will see repeatedly throughout this book), tinkering with an embryo can result in bizarre animals whose parts nonetheless develop and function in an integrative manner.[68]

Throughout the world, members of diverse societies—in pursuit of their own idiosyncratic ideals of beauty—have come to understand which features of human anatomy can and cannot easily be manipulated.[69] On the one hand, the body's soft parts—lips and ear lobes, for example—can be radically distorted, almost at will. Consider the enormous plates that women of the Surma and Bumi tribes of Ethiopia and Sudan are able to insert into their lower or upper lips. Such fantastic exaggeration of the lips, involving soft tissue alone, can be produced at any age.

On the other hand, manipulating the hard parts of the body—that is, bone—requires early intervention, ideally during periods of rapid growth. For example, archeologists have documented ancient and worldwide cultural practices aimed at deforming the skull during early infancy. Many techniques have been used: In pre-Columbian Peru, wooden boards were strapped to the heads of newborns to produce a "tower skull." In the Pacific Northwest, members of the Chinook tribe would bind a supine infant to a board and then tie another board over its head at a 25-degree angle. After many months in this position, the infant would sport a pronounced sloping skull that was a mark of distinction in Chinook society. Of course, the skull has not been the only hard part targeted for change: For about 1500 years in China, young girls were often required to bind their feet to produce a stumpy form for the inexplicable pleasure of men.

In all of these cases, societies have adopted methods that modify the mechanical forces that normally shape bone development to produce almost any desired feature. Thus, just as the growing brain is an internal force that influences the size and shape of the skull, outside forces are no less able to redirect the growth of the skull (and, as a consequence, the shape of the underlying brain). Similarly, the mechanical forces of walking throughout infancy help to shape the bones of the feet, just as those bones are shaped by foot binding. Bone growth is not predetermined; rather, bones are sculpted as the internal dynamics of growth play out within the context of real-world experience. Similar processes continue to shape and reshape bones throughout our lives—after a break, for example—but again, not as dramatically as when we are young.

Like bone, our brain and the behavior it helps produce are shaped by experience. In fact, as a learning organ, the brain specializes in adaptability. In the jargon of the field, we say that it exhibits *plasticity*. But there is more to adaptive behavior than the plasticity of the brain. In fact, behavior is shaped and molded by the hard, anatomical realities of the growing body and the concrete physical world in which that body develops.

To illustrate this point, imagine lying in bed on a hot summer night. You have no air conditioning and your only source of relief is the ceiling fan. You throw off the covers to help you cool down, but still you feel hot and sweaty. If you are like most people, you sprawl out on the bed, legs and arms outstretched, as you maximize the amount of skin surface from which to lose the heat. Now imagine that it is a dreadfully cold winter night and your heating system has failed, so you are under the covers, struggling to stay warm. You curl up in a ball, your knees close to your chest, your arms held tightly to your torso. Indeed, the very thought of stretching out gives you a chill.

How do we know which postural adjustments will bring maximal relief during those hot summer days and frigid winter nights? You might wonder whether this knowledge is hardwired in our brain, but there really is no need to hardwire that which is learned quickly and efficiently through experience. Thus, as our bodies grow, as we fatten up and slim down, and as we experience a wide range of temperatures, we learn how best to maintain thermal comfort. We buy special clothes—Bermuda shorts or a thick woolen parka. We cool or heat our home. We drink iced or hot tea. We learn to adjust our posture. The ideal positioning of our torso and limbs may be slightly different for each of us, but we quickly learn that sprawling works in the heat and curling up works in the cold. We learn for ourselves *and* we learn similar things. For these reasons, our behavioral responses to hot and cold conditions are both personal *and* universal.

Consider a second example, one that is admittedly bizarre. Consider the fact that when humans urinate, females typically squat and males typically stand. Human cultures recognize this universal feature of human behavior and reinforce it in the design of our restrooms: urinals for men in public buildings and liftable toilet seats for them everywhere else. Such a universally expressed sex-specific behavior suggests the presence of an instinct, one that might even inspire the search for a genetic cause. But such a search would be pointless.

Of course, it is obvious that the chromosomes associated with being male and female—XY and XX—would be associated with standing or squatting to urinate. But the more important point is that no specialized behavior-altering genes are necessary. Why? Because the standard urinating position of each gender simply reflects the fact that males have an external, tubular, urination apparatus—a penis—and females do not; and because females (and males) accumulate similar life experiences with their anatomical parts, most females (and males) arrive at the same behavioral destination. Thus, females prefer to squat because the alternative is the discomfort of a moistened leg. For their part, males prefer to stand because sitting takes time and effort; in effect, males urinate while standing because they can.

The implications of these seemingly trivial examples are both serious and edifying because curling up, sprawling, and urinating are universally expressed behaviors that nonetheless can be explained without appealing to "genetic hardwiring of the brain" or other exotic and magical forces within the animal. These examples show us how behavior—like a skull squeezed between two boards—can be molded under the guiding influence of persistent and mundane external factors.

Given our world and our bodies, certain behavioral outcomes are inevitable. But any satisfactory explanation of development requires a detailed understanding of the mechanisms that give rise to anatomy and behavior. To achieve such an understanding, we must first accept the fact that the cycles and spirals of development—the loops within loops within loops—preclude a straightforward story of causes and effects. Too many of our cherished concepts, nurtured in the presumption of a linear, rational, and ordered world, simply fail to adequately capture the turmoil of development.

Anomalies reflect that turmoil. They jar our senses, challenge our complacency, and force us to confront our preconceptions; they help us appreciate the fine balance that exists between the idiosyncrasies of individual development and the vagaries of evolutionary history; and they open a window to the nature of development itself—to the moment-to-moment construction of the bodies and behaviors of all animals, be they "freaks" or "normals." Whether we are talking about two-headed sisters and two-legged goats, or armless wonders and limbless snakes, or disfigured frogs and fish that switch sex, or you and me, development is very personal.

TWO

ARRESTING FEATURES
Development is All about Time

*... the analysis of normal developmental processes and
the experimental study of monstrous development
[are] one and the same problem."*
CHARLES STOCKARD (1921)[1]

*... the speeds at which the internal factors work
are of great importance in development,
and variations in the relative speeds of the
various factors may play an important part
in the relation of ontogeny to phylogeny."*
GAVIN DE BEER (1940)[2]

For one moment, many years ago, I was a great drummer—at
least I felt like one. In any event, it was just a moment.
I had been working for days on the most complicated beat I had
ever attempted. It required each of my limbs to dance rapidly
in a temporally precise pattern that, as the beat was notated,
seemed too intricate for one body with just four limbs. But on
this day I nailed it, my hands and feet punching out the rhythm
as my mind, lagging slightly behind, observed and enjoyed the

goings-on as much as a second listener might have. From this respectful distance, my mind started to intrude, now trying to control—not just observe—the individual movements of each of my limbs. Suddenly everything fell apart, sticks and limbs colliding like a pile-up on the interstate.

Successful drumming requires well-timed effort among diverse parts: finely tuned commands from the nervous system controlling muscles that are connected to stiff bones by elastic tendons. This mechanical system is extended to sticks gripped firmly yet flexibly in each hand, and to the "skin" of each drum. Every strike of the stick against the drumhead produces a reaction that influences the timing of the next strike. Hit off center or too hard and the stick bounces off in a direction or with a speed that's unexpected, altering and occasionally disrupting the timing of subsequent movements. In my case, merely focusing on my limb movements was enough to upset the delicate timing relationships that I had achieved through hours of practice.

Development also entails a balance among diverse parts interacting through time. As we will now see, even subtle disruptions to this balance can dramatically alter the development of the face and head, resulting in the emergence of unexpected and even monstrously novel structures. But it will also become clear that these "unexpected" novelties are not randomly produced. On the contrary, as Pere Alberch observed, there is a logic to monsters. To fully appreciate that logic, we must never neglect the central importance of time.

FACE TIME

"Ever wonder where our worst nightmares come from?" So began a 2003 news report announcing the discovery of the fossilized remains of a giant and now long-extinct "one-eyed" creature on the Greek island of Crete.[3] Considered in the context of Homer's terrifying description of a man-eating Cyclops in *The Odyssey*, written over 2500 years ago, the discovery suggests a possible source for Homer's inspiration.[4] But although this fossil belonged to an elephantine relative, not to a Cyclops, the news report's allusion to a single "eye socket" implies a connection between myth and reality that is not far fetched. Consider a modern elephant's skull, such as the one shown on the following page, with a large hole in the forehead. Of course the hole is not an eye socket. Rather it marks the location of the nasal cavity to which the trunk connects. The true eye sockets sit inconspicuously off to the sides. Thus, looking at this skull head-on, we see a large central hole that an ancient Greek, unfamiliar with living elephants, could easily have mistaken for the former home of a single, centrally located eye.

Although compelling, the notion that mythological creatures emerged from human encounters with the fossils of long-extinct creatures may not be the whole story. As mentioned in Chapter 1, the iconic images of the Roman gods Janus and Atlas eerily reflect embryological, not paleontological, forms. The same may be true of Homer's Cyclops. After all, why imagine

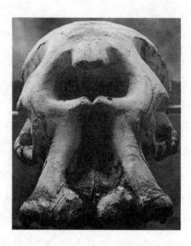

An elephant skull. Note the large central nasal cavity. The eye sockets are barely visible off to the sides.

adding flesh to bone to envision a Cyclops when some infants of humans and other animals appear in the exact form—if not size—of Homer's creation? Perhaps—and this is mere speculation—the mythical Cyclops arose by combining a mistaken interpretation of a giant fossil skull with the unmistakable horror of a tiny cyclopic infant.

Leaving aside the origins of the mythical Cyclops, we need not speculate about the origins of cyclopia because we already know a lot about it. For example, we know that cyclopia is the extreme form of a series of abnormalities that affect the entire face. These facial abnormalities mask equally extreme problems with the brain. In fact, cyclopia and its associated defects are known collectively as *holoprosencephaly*, a name that highlights the failure of the forebrain to divide into two separate halves. The incidence of holoprosencephaly may be as high as 1 in 250 fetuses, but because most of them do not survive to term, only about 1 in 16,000 infants are actually born with this condition.[5] In the most severe cases, including cyclopia, nearly all will die within one week.[6]

In the early twentieth century, the zoologist Harris Hawthorne Wilder sought to develop a conceptual framework to better understand cyclopia and related abnormalities. In particular, it impressed Wilder that "the cases usually classed as 'monstrosities' can be as natural and symmetrical in their development as are normal individuals...."[7] But it is not so easy to adopt Wilder's noble perspective while gazing upon a cyclopic infant. Wilder's perspective can come only with repeated exposure and desensitization.

For example, consider the sketches on the following page from one of Wilder's papers.[8] If this is your first experience with cyclopia, you may wonder whether such creatures are any more real than the most fanciful beasts of human imagination. At first, you may find that the single eye grips your attention. But over time, you may begin to notice the empty space between eye and mouth and even perhaps that odd structure above the eye (at first, I thought this structure was only a smudge on the page). Would you, like Wilder, choose to describe these images as "natural"?

Because my introduction to cyclopia came through a cartoon, I remained unconvinced that such infants actually exist until I saw photographic proof. But although the photographs convinced me of the reality of cyclopia, each image—of chemically preserved infants staring with that single unblinking eye—only pushed me further away from the lofty perspective that

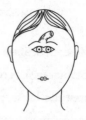

Two sketches from Wilder's "Cosmobia" series depicting two
forms of cyclopia. Note the absence of a nose in the middle
of the face and the proboscis above the eye.

Wilder espoused. The cartoons, on the other hand, provided much-needed distance such that, over time, I gradually came to appreciate cyclopia as a biological form worthy of careful attention.

As I became accustomed to the presence of a single, centrally located eye and as I studied Wilder's other sketches, one question became increasingly salient and puzzling to me. What, I wondered, might account for the fact that some of the infants Wilder examined had noses and some did not?

Today, scientists inundate us with "discoveries" of genes for every identifiable trait, such as depression, thrill seeking, and jealousy. Well, within the realm of identifiable human traits, the nose is certainly more distinct than depression or thrill seeking. So, does its absence in a cyclopic infant imply that the "gene for the nose" has wandered off with the "gene for two eyes"? Are these two genes somehow linked, like conjoined twins? Or perhaps these two genes comprise a single fragment of DNA that serves two facial functions.

To entertain these simplistic genetic fantasies is to take our eyes off the developmental processes that truly matter. Wilder revealed these processes by creating and ordering a series of images (not unlike a geologist conveying the erosive power of water using time-lapse photography). Thus, the next face in Wilder's series, presented below, now shows two eyes—close-set but nonetheless distinct—and, for the first time, a nose.

The nose's appearance in its proper location seems almost refreshing, as if the world has been made right again. But where did it come from? Glance back and forth between the various sketches and the answer presents itself: That "thing" above the cyclopic eye is a nascent nose (a *proboscis* in technical jargon), its path to the middle of the face blocked by eyes that have not gotten out of the way. In Wilder's words, written a century ago, "the nose, which is prevented from coming down in the usual manner through a downward growth of the fronto-nasal process, remains above the double eye and presents a shape something like a proboscis, decidedly abnormal, but character-istic of all monsters in whom there is no space between the eye components."[9]

Two more sketches from Wilder's "Cosmobia" series depicting
a properly located but rudimentary nose (left) and a fully
formed nose (right).

But even the nose depicted here, having squeezed through the narrow space afforded by the barely separated eyes, is only "a small and narrow nose rudiment, usually with a single median nostril." When the eyes provide more space, as shown in the final sketch in this series, a normal nose results.

We now know much more than Wilder did about how the developing eyes influence the developing nose. In particular, the eyes do not begin as a single centrally located eye-field that then divides and moves laterally to produce two eyes.[10] Rather, the eye-fields begin as a continuous line of cells awaiting a chemical trigger that suppresses activity in the eye field's central region. When this suppression occurs, the line of cells is divided into two separate eye-fields that can then develop further into two properly placed eyes.

Like cars arriving at a crowded intersection, the group of cells that will produce a nose cannot move along until the eye fields have given way. The nose cannot go around, and so it must wait for an opening and that opening must arrive on time. Otherwise, the face will be stuck with that oddly shaped and positioned proboscis that signifies the cyclopic infant.

The fact that the same cells that build a nose can so easily produce a tubular proboscis reflects a fundamental developmental process—called *induction*—whereby interactions between neighboring tissues stimulate changes in gene activity and produce new tissue. In other words, body parts arise when cells of one type interact with neighboring cells of another type, with the *interaction* producing gene activity that *induces* the development

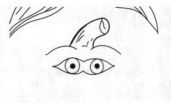

Enlargement of Wilder's sketch of a cyclopic infant to show the proboscis located above the eye (left). A young elephant (right).

of the body part in question. Aquiline nose here—ungainly proboscis over there.

Does the shape of the proboscis of the cyclopic infant remind you of the trunk of an elephant? If so, perhaps—and here I am really speculating—elephants evolved trunks by manipulating the developmental interactions among the cells that produce eyes and proboscis. This speculation seems a bit less fanciful when we compare the fetus of a normal elephant[11] with that of a cyclopic mouse,[12] as shown on the following page. Now the striking resemblance between their enlarged proboscises is even more apparent. Thus, not only does the nasal cavity of an elephant resemble the bony socket for a cyclopic eye, but the fleshy trunk that connects to the nasal cavity also resembles the nose of a cyclopic fetus. Whereas the former resemblance may be superficial, the latter may reflect a deep embryological link.

Although this embryological link between the proboscises of elephants and cyclopic infants is speculative, the lesson illustrated by Wilder's sketches is not: The same tissue can grow to be a nose or a proboscis depending upon the nature of the local cellular interactions. In the case of the nose, avoiding the

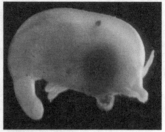

An African elephant fetus fifty-eight days after conception (left).
An embryonic mouse engineered without the sonic hedgehog
gene (right); note the enlarged proboscis.

development of an unsightly proboscis depends upon the *timely* movement of cells from one locale to another.

———

AROUND THE SAME time as Homer, 7,000 miles west of Crete in Tlatilco, a small village on the outskirts of present-day Mexico City, sculptors were representing a different category of facial disfigurement. Their striking creations would mesmerize the archaeologists who unearthed them 3,000 years later. These small ceramic figurines depict gracefully rendered female forms. Most are naked, with bulging thighs, slim waists, blunted arms, and heads adorned with fancy hair-dos. Each face is fully realized, often giving the impression of a mask concealing the true face underneath.

A few dozen of the figurines offer much more than mere stylized representations of the female form. These figurines are typical in every respect but one: the presence of two faces occupying the single head. Scholars have usually interpreted such sculptures as representations of mythological or spiritual beings. Indeed, why would we think anything different about these two-faced Neolithic figurines?

Tlatilco figurine depicting diprosopia, in this case characterized by two mouths, two noses, and three eyes.

To answer this question, we must compare additional figurines. When Gordon Bendersky, a physician and medical historian, made such a comparison, he noted that the figurines ranged from two overlapping faces sharing a single central eye socket through near-total facial separation with the emergence of four distinct eyes.[13] Such a series might easily be mistaken for artistic variations, but it appears that these ancient sculptors had more than art in mind. As Bendersky notes, the clinical precision of the Tlatilco figurines may distinguish their makers as among the first medical illustrators known to history.

The condition modeled by these sculptors is *diprosopia* (pronounced di-PRO-so-pia), a rare congenital malformation that entails duplication of facial features. As the subsequent illustration of a woman from a medical textbook attests,[14] diprosopia does not necessarily lead to early death, although individuals with more extreme conditions, like the kitten in the figure, typically do not live long.

A woman (left) and kitten (right) with diprosopia. The kitten exhibits a more extreme form of this condition.

In an odd parallel of history that transcends 3,000 years of fascination with congenital malformation, the Tlatilco figurines bring us back to Wilder, who discussed diprosopia in the same paper in which he discussed cyclopia. Remarkably, despite a formidable temporal gap, a link between Tlatilco and Wilder binds this narrative together: The gradual transitions in the Tlatilco figurines convinced Bendersky of their real-world origins, just as the gradual transitions in biological forms helped Wilder to appreciate the embryological forces that, like a subterranean river, unite the surface features.

Now let's look at Wilder's complete series, presented on the following page. It is anchored on one end by cyclopic monsters with missing features—*monstra in defectu*—and, on the other, by diprosopic monsters with an excess of features—*monstra in excessu*. Struck by the neat arrangement of this series, he coined a term, *Cosmobia*, from the Greek, meaning *orderly living beings,* to capture his perspective on the developmental origins of monsters.

According to this perspective, cyclopia and diprosopia are not distinct conditions, but rather aspects of a single condition that occupy opposite ends of a continuous spectrum. Somewhere in the middle, around plates VII and VIII, the range of "normalcy" is contained, almost lost in a sea of diversity. Within this broad context, normalcy can seem mundane, its form no more or less miraculous, or special, or more perfect than the others. Wilder was adamant on this last point, taking issue with the notion that cyclopia and diprosopia are deformations. "Abnormal they certainly are in the sense of not being the usual form in which a given species manifests itself," Wilder wrote, "but they are not deformed."[15]

As Wilder acknowledged, he was not the first to recognize the "imperceptible gradations" that comprise a series of monsters.[16] Rather, his unique contribution was to note the "symmetrical anomalies on either side of a normal being"[17] that connect the previously disconnected *monstra in defectu* with *monstra in excessu*. In his *Theory of Cosmobia*, he inferred "a cause existing in the germ, or applied during the very early stages of development"[18] through which the full range of monsters could be created artificially.

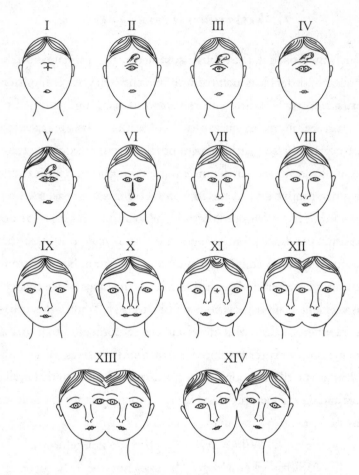

Wilder's complete "Cosmobia" series, from cyclopia (I) through normalcy
(VII) through diprosopia (X–XI). Beyond diprosopia lies a condition known
as dicephalus. Contrary to what this series suggests, dicephalus is not a
further extension of diprosopia, but rather a variant of conjoined twinning.

By *germ*, Wilder meant the genetic material contained within
sperm and egg. It was his belief that "the ultimate cause of the
development of the organism and its architectural details lies
in the germ."[19] So it followed that "true excessive or defective
Cosmobia can be produced experimentally only through a cause

which is applied early enough to lead to the formation of a germ that differs as much from the normal type of germ as the resulting organism differs from the normal adult."[20]

In 1908, as the 40-year-old Wilder wrote these words, he was corresponding with Charles Stockard, a scientist ten years his junior who had different ideas about the developmental origins of monsters. At the end of Wilder's paper, in a section devoted to recent work in the field, he admitted that Stockard's experiments with minnow and trout embryos "quite forbid me from taking a strong view concerning a germinal variation as always the necessary cause which I might otherwise have done."[21] Thus, Wilder was struggling to understand organisms as products of complex developmental cascades rather than as revelations of genetic predestination. This struggle would define in even more dramatic fashion the career of Charles Stockard and, indeed, the chaotic century that, in 1908, was just beginning.

CHARLES STOCKARD'S CENTURY

When I think back on my graduate school research at the University of Chicago, the image that comes most readily to mind is my sitting alone in a room late into the night and watching two rats, bathed in red light,[22] having sex. I know how strange that sounds, but there are more people watching rats have sex than you might imagine. In fact, the scientific study of rat sexual behavior has provided invaluable insights into the hormonal and neural mechanisms that underlie complex behavior. The great successes in this area of research include the discovery of how hormones control the estrus cycle, prepare the reproductive system for fertilization, and act on the nervous system to promote those behaviors that make reproduction possible. For my graduate work, I needed to identify female rats in heat, a state that occurs every four or five days. To identify such females, I routinely inspected vaginal cells, extracted using a swab, under a microscope.

That a vaginal swab could reveal an animal's reproductive state was discovered by George Papanicolaou, a 34-year-old assistant to Charles Stockard at Cornell University Medical College in New York City. Working with guinea pigs, Stockard needed to know when his females were in heat. Papanicolaou had the idea that monitoring vaginal cells might provide useful information quickly and efficiently.

As is so often the case in science, the value of Papanicolaou's idea projected far beyond Stockard's relatively narrow concerns. Years later, he was to make the serendipitous observation that the vaginal cells of some women are abnormal. He surmised that the presence of such abnormal cells predicts cervical cancer, still one of the leading causes of death among women. Working against an incredulous medical community, he was finally able to publish his findings in a major professional journal in 1941. Papanicolaou's method, originally devised to help Stockard identify guinea pigs in heat, is now known as the Pap smear.

But back in 1910, Stockard's research had nothing to do with cervical cancer or sexual behavior. Rather, he was interested in the effects of alcohol intoxication on the offspring of pregnant guinea pigs. These were heady times for alcohol research: The temperance movement would soon achieve its greatest victory in the United States with the 1919 ratification of the Constitution's Eighteenth Amendment, prohibiting the manufacture, sale, or transportation of alcoholic beverages. Aligned with this effort were eugenics proponents, who sought to uplift the human race through selective breeding of "superior" individuals and, when necessary, the sterilization of "degenerates." Eugenicists believed that alcoholics were corrupting the human race by passing on their inferior traits to future generations.

So when Stockard showed that prolonged and repeated alcohol intoxication in guinea pigs produced gross malformations in offspring—including complete absence of eyes—that could be passed down to the next generation, eugenicists immediately embraced his findings.[23] Stockard so welcomed their embrace that he framed his work on intoxicated guinea pigs and their "weak" offspring as the study of "racial degeneration."[24]

An immediate challenge to Stockard's conclusions came from a young geneticist, Raymond Pearl. Like Stockard, Pearl focused on the effects of alcohol intoxication, but using chickens instead of guinea pigs. He exposed male and female chickens to alcohol for one hour each day from birth to adulthood, at which time they were mated. Although alcohol treatment lowered fertility rates, Pearl found that those offspring that were produced were heartier and less likely to be deformed in comparison to the offspring of untreated parents. He attributed his findings to the detrimental effects of alcohol *only* on those eggs and sperm that were already weak—strong eggs and sperm were unaffected. Pearl further argued that although some of the weak and alcohol-exposed eggs and sperm produced deformed offspring, the predominant effect on them was lowered fertility. Thus, Pearl argued, alcohol exposure weeded out the inferior embryos, leaving behind the hardiest survivors. For Pearl, this Darwinian explanation resolved the "apparent contradiction" between his results and those of Stockard.[25]

Beyond the particulars of laboratory science, Pearl was convinced that the perceived social and political consequences of Stockard's findings flew in the face of common sense. After all, Pearl noted, the affluent and the intellectual are no strangers to the pleasures of alcohol. So if Stockard's findings were relevant to the human condition, as so many contended, Pearl—as a "hearty drinker" and a member of the educated New England elite[26]—could "see no escape from the further conclusion that a great majority of the individuals belonging to the higher intellectual and social classes in the countries of Western Europe today ought to be blind, dwarfed, and degenerate wretches...."[27] In contrast, Stockard's embryological perspective convinced him of alcohol's detrimental effects on the embryo. But with the conflict between Stockard and Pearl resting so heavily on the issue of racial degeneracy and social policy, it was easy to lose sight of the basic scientific issues that had not yet been resolved.

Of course, today we are acutely aware that alcohol is a powerful teratogen that, when consumed during pregnancy, causes (among other things) the death of rapidly dividing cells in the developing embryo. Depending on the quantity and timing of alcohol consumption, a variety of malformations and neurological problems, now collectively referred to as *fetal alcohol syndrome*, can result. Unfortunately, confirmation of alcohol's teratogenic effects would be delayed many decades, in part because Prohibition, by removing alcohol from the public sphere, ameliorated concerns about the negative consequences of alcohol consumption.[28]

Incredibly, in a turnabout, Stockard reassessed his own research and embraced Pearl's Darwinian explanation. He then promoted alcohol as a tool for achieving eugenic goals. Although once an advocate of efforts to improve prenatal care, he now suggested—before a stunned audience of temperance advocates in 1920—that alcohol, "if used in a eugenic way," could "eliminate bad individuals" by preventing them from being born.[29] Pearl was appalled and, in 1927, would distill his disdain for the eugenics movement in a remarkable article published in *The American Mercury*. In one particularly powerful passage, he described eugenics as "a mingled mess of ill-grounded and uncritical sociology, economics, anthropology, and politics, full of emotional appeals to class and race prejudices, solemnly put forth as science, and unfortunately accepted as such by the general public."[30]

As we now know, despite the efforts of Pearl and many others, the eugenics movement would not be extinguished with words. Over the next decade, Germany would carry the banner of eugenics with pride, creating sterilization programs designed to bar the mentally and physically handicapped from a future that the Nazis had reserved only for superior beings.

Germany was not alone. In 1937, at a conference organized jointly by the American Eugenics Society and the New York Academy of Medicine, the physicians in attendance discussed, with some urgency, the need for sterilization, contraception, and selective breeding as countermeasures lest society, as one attendee commented, "deteriorate due to the improper distri-bution of births…."[31]

In attendance at that conference was Charles Stockard, now a distinguished senior scientist and newly appointed President of the Board of Scientific Directors of the pro-eugenics Rockefeller Institute for Medical Research (later renamed The Rockefeller University). Stockard reportedly communicated his concerns in this way:

> If we are going to continue to live in this world and not face ultimate extermination, we must give thought to the effect of the artificial conditions of civilization, as they affect human breeding. Ultimately, propagation should be absolutely pre-vented among low grade and defective stocks who are unable to pull their own weight in the social organization. Not only is there statistical certainty that they will produce offspring

who, on the average, will have similar hereditary limitations, but, in addition, they provide a home environment which makes proper development impossible. We have never been willing to face these questions in a large enough way....[32]

In two years, Stockard would be dead. Soon thereafter in Berlin, his words, and those of his eugenicist colleagues, would be put into action in the form of a newly authorized program to exert ultimate control over hundreds of thousands of troubled people deemed unworthy of life. For this program—housed in a villa at 4 Tiergartenstrasse—sterilization did not go far enough. The T-4 *euthanasia* program was born.

Was Charles Stockard evil? In light of his active support for eugenics and the tragic legacy of that movement, one might reasonably conclude that he was. But it may be more useful to view Stockard's mission as emblematic of a troubled period in our political, scientific, and social history. In this one man, we can glimpse a century of conflict and confusion regarding such core concepts as genes, inheritance, and environment. In him, we see a scientist who began his career studying the effects of the environment on individual development, and ended it as a rabid proponent of eugenics, devoting his time to an ambitious examination of the inheritance of form and behavior in pure-bred and hybrid dogs (which he housed at his specially built farm near Peekskill, New York).[33]

Ironically, the young Stockard's challenge to Wilder prevented that elder scientist from taking "a strong view" on the role of genes in the production of cyclopia and other malformations. This is the Stockard to whom we will soon return. But before we do, let's briefly examine the efforts of those early scientists who, like the young Stockard, sought to alter embryonic development within the confines of the laboratory.

DEVIATING DEVELOPMENT

The efforts of Etienne Geoffrey Saint-Hilaire to classify the world of monsters, described in Chapter 1, were followed in 1820 by his experiments aimed at understanding the developmental processes that produce the normal and the monstrous.[34] Skeptical of the *preformationist* notion that development unfolds according to a preordained plan that is unaffected by developmental conditions, Etienne sought to "deviate development" by altering the incubation environment of chicken eggs.[35] These experiments never yielded definitive results. Similar efforts by his son, Isidore, were even less successful.

In 1860, forty years after Etienne began his experiments, the Academy of Sciences in Paris announced the Prix Alhumbert. This prize was to be awarded to the best paper demonstrating environmental modification of a developing embryo. With Isidore sitting as a member of the commission that created this competition, two awardees were announced in 1862. One of them, Camille Dareste, is now regarded as the founder of experimental teratology (though Dareste himself bestowed that distinction on Etienne).

Dareste, like Etienne, manipulated the incubation environment of chick embryos. He heated, cooled, shook, and chemically treated his eggs. Although his methods were neither perfect nor novel, he was more systematic and successful than the Saint-Hilaires and the scientist with whom he shared the Prix Alhumbert. Dareste found that his manipulations produced the severest abnormalities when he applied them early in embryonic development. He inferred that his treatments succeeded in producing malformations because they somehow slowed or arrested the process of development.

Dareste's observations and inferences make clear that from its inception, experimental teratology was intimately linked with the most basic issues of normal embryology. Nonetheless, teratology and embryology would develop as separate fields, a separation that is unnatural. Writing more than a century after Dareste claimed his prize, one embryologist described the situation like this:

> If an experiment is performed on an embryo, and the embryo nonetheless develops normally, the investigator believes he is an embryologist studying regulation. If the embryo fails to regulate and develops abnormally, if something overt goes wrong, he is studying abnormal development and he is a teratologist. An embryologist uses abnormal development as a key to the normal; an experimental teratologist tries to see where normal development went wrong and why, especially if he has clinical inclinations. But since whatever an experimental investigator does to an embryo (and in many cases

the experimentalist may be Nature) some things go right and some go wrong, all distinctions between embryology and experimental teratology become blurred; the disciplines are symbiotic.[36]

This symbiosis is evident in the seamless transition from the work of Dareste to that of Charles Stockard.

PRECIOUS MOMENTS

In 1907, Stockard reported the first in a series of methodical studies using the eggs of minnows and trout that would lead him to the same inference as Dareste's: that his experimental treatments somehow slowed or arrested the process of normal embryonic development. Moreover, Stockard's work would allow him to pinpoint those developmental moments when embryos gain the capacity to become monsters.

As with Wilder's Cosmobia series, Stockard's monsters comprise a range of deformities from the single to the double—from cyclopia to two-headedness. That such similar series exist for both human and minnow attests to the power of experimental embryology to probe the most fundamental biological processes. This power derives from a simple yet profound fact of embryology: Widely divergent forms of life share the earliest phases of development. Thus, as we face the striking developmental parallels between two evolutionarily distant animals, our attention is diverted toward those earliest embryonic moments after the egg has been fertilized, when the embryo is still overtly indistinguishable as a human or a minnow. At this time, the eyes, head, and tail are little more than fields of cells, knocking into one another, dividing, folding, gradually becoming recognizable parts.

The once-popular preformationist view of development, which had motivated Etienne Geoffrey Saint-Hilaire's foray into experimental work, imagined that the fertilized egg was, like nested Russian dolls, simply a diminutive version of the final organism into which it would grow. In stark contrast, the alternative, epigenetic view appreciates that

> the egg's organization suffices only for its development to the next immediate developmental stage, which must then make additional instructive materials to develop into a further stage, which must in turn repeat the process.... In this view, development is a series of generative processes, each building on the organization of the previous stage.[37]

Monsters help us to appreciate the critical differences between these two perspectives. How? If an embryo unfolds into a monster, then a preformationist has little choice but to imagine that the monster existed in nascent embryonic form from the outset or, at the very least, resulted from an extreme deformation of the early embryo. But if we view development epigenetically as a step-by-step, cascading series of generative processes, then a monster can be neatly understood as the destination produced when an embryo takes an alternate path at some time *during* development. There is no need for prestructure, preform, or pre-design; such static notions leave no space for *time*.

But time matters. As Stockard demonstrated in a variety of ways, the production of cyclopic and two-headed minnows is limited to a narrow window of opportunity—he called them "moments of supremacy"; today, we generally refer to them as *sensitive* or *critical periods*. For example, when Stockard exposed embryonic minnows to cool temperatures, he produced monsters of all kinds—but only if the exposure occurred within the first twenty-four hours after fertilization. In contrast, if Stockard waited more than twenty-four hours, he encountered what he called a "moment of indifference." Nothing remarkable happened.

What might be going on within the embryo that would distinguish a moment of indifference from a moment of supremacy? Building on Dareste's linkage between developmental arrest and the production of monsters, Stockard suggested that this linkage is particularly strong during a moment of supremacy because it is a time "when certain important developmental steps are in rapid progress or are just ready to enter upon rapid changes, a moment when a particular part is developing at a rate much in excess of the rate of other parts in general."[38] Accordingly, the nature of the insult—cooling, oxygen deprivation, and so on—is less important than the effect of that insult on the developmental rate. If development is slowed at the right moment, there will be monsters.

Stockard's idea that development is most modifiable when parts are changing rapidly is similar to the notion that bones are most easily deformable when growth is rapid. As with bones, which grow rapidly in brief spurts, Stockard pointed out that "eggs develop with rhythmical changes in rate,"[39] with periods of relative quiet interspersed with brief moments of turbulent transformation.

Stockard placed great emphasis upon the rhythm and rate of development. Imagine a car race, with each car representing a different embryonic cell or group of cells, and the speed of the cars representing the rate of development. If all of the cars are traveling at the same slow speed, then instructing all of them to slow down or even stop will have a negligible effect on their order in the race. Moreover, when the cars are allowed to speed up again, the previous relationships among the cars can be easily recovered.

But now imagine that the cars are traveling at different speeds. If they are suddenly commanded to stop, it will be nearly impossible to tell which car was where at the moment that the command was issued; and because each car will slow down at a different rate, cars will be passing each other even as they slow down, producing even more confusion about the relationship among the cars before the stop order was given. In a similar vein, cooling an embryo as it is undergoing a period of rapid transformation is more likely to disrupt normal development than if the same disruption is performed during a period of relative calm.

After fertilization, the first such transformative period is *gastrulation,* a process in which the newly fertilized egg begins to form discernible layers that will eventually develop into skin, gut, and brain. In addition, the primary axes of the body are established at this time, including the distinction between front and back, and top and bottom. In short, gastrulation is a moment of supremacy—of profound reorganization—of rapid and complex change. It is during gastrulation that embryonic development is most easily disrupted by the kinds of manipulations that Stockard used. Although the result of these manipulations to the embryos is often death, many of those embryos that survive develop abnormally, exhibiting a variety of malformations that includes cyclopia and twinning.

So we see that numerous environmental manipulations, applied early enough in development, are capable of producing a diversity of monstrous forms. This diversity was still another point of emphasis for Stockard. He noted his creation of "double monsters of varying degrees, from separate twins, fused but with complete bodies and tails, to double bodies and single tails, and finally different degrees of double headedness on single bodies. There are specimens exhibiting...cyclopia, and all types of malformed eyes...."[40] But this observation did not imply to Stockard the random emergence of defects. On the contrary, he found that he could disrupt the development of any particular organ by precisely timing the environmental insult:

> The localized anomaly may involve only the eye, only the bilaterality of the brain, only the ear, only the mouth structures,

only the kidneys…, only the genitalia, etc. It is evident that such anomalies could not occur unless there was a certain moment of specific and peculiar susceptibility *on the part of each organ* during which any unfavorable condition would act on it in a selective way.[41]

Moreover, because proliferation of particular organs continues beyond embryonic development, specific moments of susceptibility continue into postnatal life as well.

All of this makes clear once again how our understanding of monstrous development is intimately connected with the intricacies of normal development. It is because of this intimate connection that manufacturing monsters is surprisingly easy.

———

JUST AS WILDER commented on Stockard in his paper about monsters, so too did Stockard comment on Wilder. The elder scientist, Stockard wrote, was "misled" about the cause of monsters, thinking it was "more probable that orderly deviations from the normal would arise in the germ-plasm than that they should occur as a result of some modification during individual development."[42]

Wilder was wrong to think that monsters could arise only via a genetic mechanism, encoded in egg or sperm; as Stockard demonstrated, environmental factors can reliably alter the course of development to produce monsters. Still, Wilder was not completely wrong: for example, as is now known, genetic mutations underlie some cases of cyclopia, especially those that run in families.[43] Moreover, we should not forget that even environmental factors can produce their effects by modifying the activity of genes or the action of their products.

In other words, both Wilder and Stockard were right and wrong, their disagreement reflecting an either–or, dichotomous mentality concerning the developmental roles of genes and environment. This mentality continues to confuse many people to this day.

But this confusion evaporates by reorienting our thinking. The key is to appreciate that development arises through a network of genetic and nongenetic interactions cascading through time. Within that network, developmental events that rely on a particular gene in one instance can occur through environmental influences in another. As we will examine further in Chapter 5, sex chromosomes are absent in some animals—for example, turtles and crocodiles—but this does not prevent them from developing into males or females.

In such species, incubation temperature replaces the need for sex chromosomes: We say that the effect of temperature on the developmental network is *interchangeable*[44] with the genetic mechanism that triggers the same process in, for example, humans and dogs. Similarly, cyclopia and its related conditions—again, known collectively as holoprosencephaly—can arise through either a genetic mutation or an environmental disturbance (for example, if the mother has diabetes or consumes alcohol), but in either case, the same developmental network is being modified. Clearly, we would be wrong to label *all* infants with holoprosencephaly as *mutants*.

Since Stockard's time, we have learned a lot about holo-prosencephaly and the network of mechanisms that produces it. Recall that the name itself refers not to a malformed face, but to the failure of the forebrain to divide into two symmetri-cal halves. In fact, there is considerable variability in the degree to which the forebrain fails to divide, resulting in related var-iability in brain function and survivability in these infants.[45] Similarly, the faces of infants with holoprosencephaly can display cyclopia as well as a variety of other malformations of the eyes, nose, and mouth (including cleft palate). Thus, one of the characteristics of holoprosencephaly—even in those instances where it runs in families with known genetic mutations—is its variability.

The search for the mechanisms that produce holoprosen-cephaly in all its forms took its biggest step forward in the 1960s as a result of a cyclopia epidemic among sheep in Utah. The culprit was soon identified: Pregnant ewes were eating a highly toxic range plant known as the false hellebore, *Veratrum californicum*.[46] Subsequent research indicated that ingestion of the plant on the fourteenth day of gestation—that is, around the time of gastrulation, as Stockard would have predicted—was necessary to induce cyclopia. The critical ingredient of the plant was also soon identified and given the name, cyclopamine.

Cyclopamine produces its horrible effects in sheep, as well as goats, rabbits, hamsters, and chickens, so long as it reaches the embryonic environment around the time of gastrulation.[47] To affect so many species in such a predictable way suggested that cyclopamine was interfering with a fundamental developmental mechanism. This is now known to be the case. As was recently discovered,[48] cyclopamine interferes with the function of a protein called sonic hedgehog (Shh)—the protein that results when the sonic hedgehog gene (*Shh*) is expressed.[49]

Among developmental biologists, *Shh* is a superstar, designated by Evo Devo enthusiasts as one of the "master genes."[50] Although much of the attention that *Shh* has received is deserved, it is easy to get too caught up in all the hoopla. Yes, identifying a relatively small set of genetic components involved in numerous developmental processes has provided important insights into both development and evolution, and *Shh* is undoubtedly a central player in these revelations. But *Shh* is only one component in a complex and dynamic machine. Thus, although mutations of *Shh* can produce holoprosencephaly, there are many other separate mutations that can produce the same condition. Moreover, although the detrimental effects of cyclopamine are mediated through the Shh system, there are other environmental factors that can have similar effects by acting on other components of the developmental network.

One simple, stubborn fact remains: No single genetic or environmental factor is able to account for the diverse malformations exhibited by infants with holoprosencephaly. Even infants with identical Shh mutations can exhibit the full range of malformations associated with the condition. How can this be? As one group of investigators recently noted, this is "one of the most perplexing questions in clinical genetics...."[51] After all, should we not expect greater control from a gene that has attained the elevated status of "master"?

Our path out of this perplexity brings us back to the central importance of time. *Shh* is not special because of its linkage to particular traits, but rather because of its participation in a variety of similar processes. These processes occur at various locations within the embryo and at various times throughout development. For example, in the head, *Shh* is activated sequentially (and thus its associated protein, Shh, is produced sequentially), beginning in embryonic tissue in the brain and continuing in tissue that will form the face. One group of scientists,[52] working with chicken embryos, identified six distinct times when *Shh* is expressed and when important developmental events are occurring. Then, by exposing the embryo to cyclopamine at these times, the scientists blocked the Shh signaling system at each moment and location, and observed the outcome.

In effect, what these scientists found provides direct and more detailed support for what Stockard argued nearly a century earlier based on his work with minnows: Timing is key.[53] Exposure of chicken embryos to cyclopamine at different times produces specific malformations that reflect the particular region of the brain or face expressing *Shh* at the moment of exposure.

Accordingly, because *Shh* is expressed around the time of gastrulation when the nervous system is only beginning to develop, cyclopamine produces its most dramatic results when given around this time. If the embryo survives, it exhibits a fused mass of brain tissue. Also, because development of the brain acts as a scaffold upon which the face is constructed, any massive malformation of the brain results secondarily in cyclopia. Wait a bit longer before giving cyclopamine—around the time that *Shh* is again being expressed, but this time specifically in the forebrain—and the result again is a mass of fused brain tissue and a range of severe facial defects (including extremely close-set eyes), but not full-blown cyclopia.

Now, wait a few hours longer—long enough that *Shh* has been expressed in the forebrain but not yet in the face—and now only the face is adversely affected. These facial defects are due to cyclopamine's effects within the face itself, rather than to secondary structural disturbances resulting from a malformed brain. Finally, give cyclopamine after *Shh* has been expressed in the face and the embryo now develops normally, even after very high doses.

Just as the downstream effects of a river dam are more severe the closer it is placed to the river's source, the effects of blocking *Shh* activity with cyclopamine become increasingly severe at earlier times in development. As the developmental process flows downstream, the effects of cyclopamine become increasingly focused to the point when it is possible to produce gross malformations of the face while completely sparing the brain. Thus, according to the authors of this study, "not only was embryonic age an important determinant of the teratogenic effect of cyclopamine but...the timing of *Shh* induction and expression appeared to be equally relevant to the severity" of the facial defects that cyclopamine produced.[54]

We can now begin to understand how a single mutation of *Shh* can produce such a wide diversity of effects on brain and face: *Shh* is involved in a variety of similar processes that, because they recur at distinct anatomical locations and times, produce distinct results. Moreover, at each moment in developmental time, events at one location can set the stage for those at another; for example, defects in brain development can deprive the face of critical structural support that it will need to develop normally. This complex epigenetic cascade of gene expression and regulation—locked in an intimate embrace with time— makes the challenge of understanding development more than a mere cataloguing of genes.

The failure of the brain to divide into two lobes—and of the eye fields to produce two separate eyes—reflects a failure of development on the embryo's midline. So it is not surprising that cells that are concentrated on the midline express *Shh*. This explains why blocking Shh signaling with cyclopamine can play havoc with the development of structures that normally divide on the midline. Now, returning to the first half of Wilder's Cosmobia series, we can see how blocking *Shh* expression can produce deficits of midline division, as with cyclopia. But what about those conditions, such as diprosopia, that are characterized by an overabundance of division and that account for the other half of Wilder's series?

Here, too, we can tap into the epigenetic cascade that includes *Shh* to provide an answer. But now, rather than block Shh signaling with cyclopamine, we increase exposure to the Shh protein and observe the effects. True to expectation, increasing the amount of Shh available to the developing embryo increases activity on the midline, thereby widening the face and brain.

Adding Shh to the developmental stew has not yet produced the additional eyes and noses that characterize diprosopia. But it is clear that the width of the face is a regulated developmental characteristic and that *Shh* is one of many genes, and Shh one of many proteins, involved in this process. As we await final determination of the mechanisms involved, we can marvel at how the basic features of brain and face are constructed from materials both found and purchased within the chemical concoction that is the developing embryo.

PAS DE DEUX

Looks can be deceiving. Recall how Wilder's Cosmobia series depicts a range of conditions: from the collapsed midline of cyclopia, through normalcy, then the widening midline of diprosopia, to dicephalus. Given Shh's role in the widening of the face, we might imagine that Shh activity (and that of other related molecules) can become so excessive that it actually unzips the embryo to produce two distinct heads. If so, perhaps the unzipping can continue to produce twins that share fewer and fewer parts until two complete but identical twins emerge.

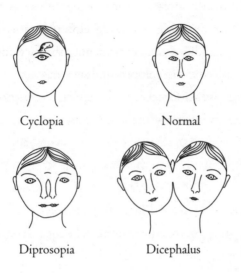

Cyclopia   Normal

Diprosopia   Dicephalus

But what I have just described is a fantasy arising from taking Wilder's Cosmobia series and its implied continuity a bit too seriously. In fact, Wilder's complete series does not depict a single process, either in time or space. As discussed, even *Shh* is expressed during development at multiple moments and in multiple locations to produce the graded facial continuum from cyclopia to diprosopia. But to explain dicephalus, the next step in Wilder's series, we must jump back in time—all the way back to the earliest moments of embryonic development. For, rather than representing the next developmental step beyond diprosopia, dicephalus is a wholly separate process deserving a wholly separate discussion of the mechanisms that produce twins—identical, fraternal, and conjoined.

The following figure displays yet another set of human infants, this one not so much a continuum as a nonexhaustive collection. The shaded figure, illustrating diprosopia, is included here only for contrast. It does not belong with the others. In fact, its inclusion here is as misleading as the inclusion of dicephalus in Wilder's Cosmobia series.

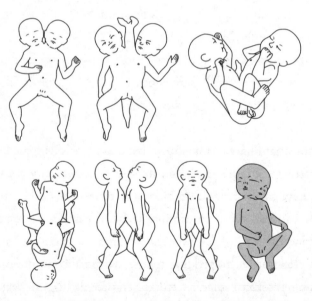

Some examples of human conjoined twinning. The shaded figure
(at lower-right) depicts diprosopus, which has a distinct
developmental origin and thus should not be included
within the category of conjoined twins.

How do we know that diprosopia does not belong within the class of conjoined twins? The answer to this question, which is so elusive that even experts have disagreed, relies on information from a variety of sources. For example, although diprosopic fetuses are more likely to be female, dicephalic fetuses are equally likely to be male or female, thus suggesting the involvement of different developmental processes.[55] Perhaps even more compelling is the fact that the face of a human fetus develops between the fourth and eighth weeks of gestation, whereas gastrulation—after which twinning can no longer occur—is completed by the end of the second week. In other words, by the time that diprosopia develops in a human fetus, the time for conjoined twinning is long past.

Conjoined twins are only the most exotic of twins.[56] The least exotic are fraternal or dizygotic twins, which are produced when two individual eggs are fertilized by two sperm. Each twin develops separately and securely within its own dedicated uterine environment, enveloped within its own amnion and chorion, or the inner- and outermost membranes, respectively, that surround the fetus. Typically, each twin also has a separate, dedicated connection to the mother's circulation through its own placenta.

In contrast with dizygotic twins, identical or monozygotic twins are produced when a fertilized egg divides to produce two independent fetuses. The type of monozygotic twin produced depends on the *timing* of the twinning event. In about one third of all cases of monozygotic twins in humans, this division occurs within three days of fertilization when the embryo comprises just two cells. Then, when the two embryos implant in the uterine wall, they do so in a fashion similar to that of dizygotic twins. That is, each fetus possesses its own amnion and chorion, as if they were two siblings living in the same house but sleeping in separate bedrooms.

But for the vast majority of monozygotic twins, the twinning event happens within four to eight days after fertilization. At this time, the embryo is in the process of implanting in the uterine wall. Under these circumstances, the resulting twin fetuses come to share a common chorion and, therefore, a common placental circulation. But each is encased within its own amnion, as if the two siblings were sharing the same bedroom but sleeping in different beds.

Very rarely, the monozygotic siblings come to share the same bed as well. In such cases, the twinning event occurs two weeks after fertilization, by which time the embryo has implanted in the uterine wall and gastrulation has occurred. This type of twinning is often associated with malformations and complications, including conjoined twinning, which results from the incomplete division of the embryo. (Although conjoined twinning is now thought by most experts to arise from the division, or fission, of a single embryo,[57] some still maintain that these twins arise from the partial rejoining, or fusion, of two previously separate twins.[58]) These twins are the rarest of the rare, roughly occurring only once in every 100,000 births. Survival is even more rare: Most die within one day of birth.

The most common types of conjoined twins are joined at the chest and abdomen. Some are joined side-to-side (as in dicephalus), others head-to-head or rump-to-rump. Occasionally, one of the twins is malformed, producing an asymmetrical relationship between the two. For example, if one twin lacks a functioning heart, the malformed twin will feed parasitically off the circulation of the other. In some extraordinary cases, as with Laloo and his parasitic twin, described in Chapter 1, one of the twins can grow partially inside the other.

Even more bizarre is *fetus in fetu*, in which one twin grows completely inside the other. In a recent and particularly shocking case,[59] an Indian man who had endured life with a rather large belly found, at the age of thirty-six, that he was having trouble breathing. Believing he had a rapidly growing tumor that was interfering with the movement of his diaphragm, the man's doctors decided to operate. As the surgeon opened up the man's belly and searched for the tumor, he found himself shaking a well-developed hand with long fingernails. Then came more limbs, bones, hair, genitalia—a jumble of body parts that, under different circumstances, would have cohered to form this man's twin brother.

This man lived with a twin he never knew, an odd situation to be sure but one that pales in comparison to the enduring challenges faced by conjoined twins as they struggle to make a life together. Nowhere are these challenges more apparent than with the dicephalic twins Abigail and Brittany Hensel, introduced earlier in this book. Born in Minnesota in 1990, they are among only a handful of such twins in recorded history to survive with two heads, two spinal cords, two hearts, but only one pair of legs. Each twin primarily controls her own arm. (A third arm, which emerged awkwardly from their shared middle shoulder, was removed soon after birth.) Nonetheless, the twins have learned to coordinate their activities, working jointly to play the piano and shuffle a deck of cards. Such skills have arisen through lifelong learning experiences, just as jazz musicians learn to anticipate and respond to each other's improvisational ideas.

Other aspects of the Hensel twins' behavior must involve even more intimate coordination. After all, the lower half of their body is completely shared. Thus, their two nervous systems likely share control of both legs. It is not difficult to imagine how crosstalk and confusion might result from two nervous systems vying for control over each leg like two dogs fighting over a bone.

But the twins are not rendered immobile by their predicament. Far from it. As with the bipedal goat, these remarkable girls have successfully adapted their behavior to a unique biological condition. It may be that evolution has never favored dicephalus, but evolution also has never precluded living—even thriving—with this condition.

Perhaps more than any two sisters in the world, we would expect the Hensel twins occasionally to disagree. Indeed, in a truly self-defeating act, Brittany reportedly once hit Abigail on the head with a rock.[60] It has also been reported that when the two sisters cannot agree on a destination, paralysis can set in.

Similarly, two-headed snakes have been known to fight over food even though both heads benefit from each meal consumed; and when the two heads pull in opposite directions, one must win the battle if the entire snake—including both heads—is to move forward (and typically one "dominant" head wins most of these battles).

Perhaps these glitches are what doom dicephalic creatures in the wild and prevent the evolution of a dicephalic species. Evolution is a battle waged at the margins—where even the occasional food fight or moment of indecision is enough to compromise survival. Sometimes the costs are quite severe; for example, dicephalic turtles cannot retract both heads into a protective shell that has room for just one. In addition to these obvious costs, a dicephalic species is unlikely to evolve without a clear benefit to possessing two heads.

But if an animal were ever to find itself in an environment that, for whatever reason, did favor the possession of two heads, a dicephalic species could arise very rapidly for one simple reason: *The embryo's potential to produce two heads is no less ancient, and no less fundamental, than its potential to produce just one.* So there it sits, like so many other embryonic potentials, waiting in the wings should evolution ever see fit to call it forward for active duty.

Not only has evolution failed to call dicephalus forward, it has actually shaped the developmental process to make it less likely. This simple observation highlights with particular clarity this singular fact: Like Brittany and Abigail Hensel, evolution and development are inseparable.

THE CONJOINED NATURE OF
DEVELOPMENT AND EVOLUTION

To understand the forces that regulate the development of dicephalus and other forms of twinning, let's examine the unfertilized egg—an amorphous sphere resembling a globe with contents distributed into northern and southern hemispheres. A surface "crust" encloses the chemical constituents within, like the peel of an orange enclosing the edible core. When a sperm successfully penetrates the crust—an event that occurs unpredictably at a random location in the egg's northern hemisphere—it leaves its mark like a flag planted in a new world. But even more important, fertilization sets in motion a series of chemical rearrangements and surface rotations that, in short order, establish the precursors to the nervous system and vertebral column. At first only visible as a darkened streak running from pole to pole on the embryo's surface, the incipient vertebral column presents the first visible evidence that a single individual is on its way. In frogs, this individual is a tadpole, hatched a mere 48 hours after fertilization.

Sometimes two streaks arise. When this happens, we know that a twinning event has occurred. In frogs, these twins are typically conjoined, presenting two heads that share a single body, not unlike the Hensels. Such twins in frogs have proved of great value to experimental embryologists seeking to identify the critical events around the time of gastrulation that trigger the nervous system's formation. By demonstrating how to produce two, we learn how to produce just one.

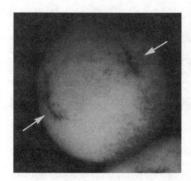 

The fertilized egg of a frog embryo can be artificially stimulated to produce two streaks (left, indicated by arrows), eventually resulting in conjoined twin tadpoles with two heads (right).

What, then, determines twinning? In frogs, scientists have taken note of the rotation of the embryonic crust soon after fertilization, as if the orange's peel were rotating independently of the core. The key to resolving the mystery of twinning was to understand how the rotation of the crust carries along a chemical determinant from one location to another, thereby bringing this determinant into contact with other chemical factors within the core. In this way, chemicals that were once far from each other are now brought into direct contact, instigating *inductive* interactions that trigger the specification of the embryo's *primary axis*, that is, the formation of the nervous system and vertebral column.

Manipulations that alter the rotation of the embryonic crust around the core can trigger the formation of a *secondary axis*, that is, a twin. Indeed, in the mid-1980s, scientists refined a technique for creating tadpole twins that involved spinning newly fertilized eggs in a centrifuge.[61] However, not any kind of spinning would do. Like bakers refining the recipe for a cake, these scientists found that twins could be reliably produced when embryos experienced a specific centrifugal force (thirty times the force of gravity), for a specific duration (four minutes), and at a specific orientation in relation to the centrifugal force (ninety degrees).

The scientists also found that timing was a critical factor: Eggs that had passed a little less than half the temporal distance between insemination and first cleavage (that is, the time when the single-cell zygote divides into two cells) were much more likely to twin. Subsequent and painstaking research has aimed to identify the critical chemical interactions in the core and crust that trigger the formation of a dual embryo.

We have seen how Charles Stockard, through careful manipulation of the developmental environment, was able to produce minnows and trout with varying degrees of conjoined twinning. In his laboratory, he found that changes in temperature and oxygen supply were effective for producing twins. But outside the laboratory, the true relevance of all this research reveals itself. After all, the natural world is much like a laboratory—though without the control.

For example, dicephalus is relatively common among reptiles because developing eggs are exposed to fluctuating environmental conditions without the protection and care of the mother that laid them. Consider the British grass snake, of which hundreds of dicephalic creatures have been reported. What makes this species so interesting is that its eggs are typically laid inside compost heaps, within which decomposing organic matter produces heat. This heat is particularly valuable to a pregnant snake seeking to incubate its eggs in an intemperate British climate.

Dicephalus in a British grass snake. High incubation temperatures cause this condition. Therefore, this snake may be a freak, but it is not a mutant.

But a compost heap is the developmental equivalent of a crapshoot. No one—certainly not a snake—can predict or control the amount of heat produced. So when temperatures rise above forty degrees Celsius, most of the eggs will simply fail to hatch. A small percentage of those eggs that do hatch will emerge with two heads. This is one of the risks of incubating eggs outside the mother's body.

Twinning is rare in birds, especially compared to reptiles. But among birds, ducks produce a relatively high percentage of double monsters (about two percent).[62] Why ducks? The answer appears related not to temperature, but to the rotations experienced by the embryo during early development. Recall how conjoined tadpole twins can be produced experimentally by spinning an egg during a sensitive period of development. Ducks have a large, elastic uterus, in which the egg is free to roll around. It is thought that when these rolling movements happen in just the right way and at just the right time, interactions among the egg's chemical constituents are critically altered. So as with the thermal fluctuations of the compost heap, the duck's uterus appears to provide a natural environment conducive to the production of conjoined twins.

Although reptiles must surrender their eggs to the relative unpredictability of the external environment, birds improve their odds through active participation in the process of incubation. From laying to hatching, mother and father provide warmth and protection. But such continuous parental care masks a momentous discontinuity: In its journey from uterus to nest, the egg must endure a sudden and dramatic change in temperature. Based on everything we have discussed so far, such an event could easily provoke the kinds of embryonic rearrangements that Stockard and others produced experimentally in the laboratory. So why don't we inhabit a world of two-headed birds?

The answer is simple yet profound: To prevent the regular development of double monsters, birds have evolved so that the thermal shock that accompanies the laying of an egg occurs only after the embryo has passed through the most sensitive period of embryonic development—that is, the period that ends with gastrulation. As Stockard noted, the "important matter of a few hours' difference in egg-laying time lies between the successful class of birds and a hopelessly unfit monstrous condition."[63]

From the preceding discussion, we can begin to sense the folly of maintaining a wall of separation between development and evolution. Such a wall cannot be sustained because *development itself—including the developmental environment—evolves.* Once we appreciate the implications of this obvious yet underappreciated notion, false dichotomies fade away. In their place, a more coherent picture of both development and evolution comes into view. We see that one outcome of the evolutionary process is the avoidance of developmental conditions that produce anomalous and unfit outcomes.

Still, the eccentricities of the natural world will occasionally exceed the regulatory efforts of animals and the reach of evolutionary modification. "No developmental environment in nature is constantly perfect," Stockard observed, "and this fact is the underlying cause of the frequently occurring malformation and monstrous production."[64]

In recent years, biologists have turned increasing attention to the fact that animals construct their environments—their niches—to suit their needs.[65] Termites build elaborate mounds of mud that satisfy the needs of a complex colony; beavers build dams that permit the construction of protective shelters along creek beds; spiders weave webs to ensnare prey; naked mole-rats dig elaborate underground tunnel systems. By extension, the eggshell and the uterus are developmental environments that enable parents to produce viable offspring.

But there can be no distinct line of demarcation between the construction of a nest, dam, web, or tunnel, and the casting of a calcite shell or the physical joining of mother and fetus via the umbilical cord. Nor can we easily demarcate these forms of niche construction from the epic efforts of green turtles to navigate across hundreds of miles of open water back to the sandy beaches on which they were hatched.

As we learn more about the mechanistic details of development, we will come to see that evolution is not omnipotent: It simply cannot produce creatures that development does not allow. On the flip side, as this chapter has illustrated, the oddity of a creature provides little insight into the ease of producing it. Indeed, some of the oddest creatures are remarkably easy to produce.

Traditionally, the driving force behind such oddities—and variability in general—has been the mutation. Even scientists like Marc Kirschner and John Gerhart, who are notable for emphasizing the role that developmental mechanisms play in the evolution of novelty, still look to the mutation to get the evolutionary ball rolling:

> First, genetic variation is required for evolutionary change. Genetic variation initially arises by mutation. Much of the genetic change that is important in evolution comes from the reassortment of mutations of previous generations by sexual reproduction....Novelty in the organism's physiology, anatomy, or behaviors arises mostly by the use of conserved processes in new combinations, at different times, and in different places and amounts, rather than by the invention of new processes.[66]

Mary Jane West-Eberhard shares Kirschner and Gerhart's belief that development is a central player in the production of evolutionary novelties, but she does not share their commitment to the mutation as the sole initiating force:

> Contrary to the notion that mutational novelties have superior evolutionary potential, there are strong arguments for the greater evolutionary potential of environmentally induced novelties. An environmental factor can affect numerous individuals at once, whereas a mutation initially can affect only one.[67]

Elsewhere, she expands on this view:

> First, environmental induction is a major initiator of adaptive evolutionary change. The origin and evolution of adaptive novelty do not await mutation; on the contrary, *genes are followers, not leaders, in evolution*. Second, evolutionary novelties result from the reorganization of preexisting phenotypes and the incorporation of environmental elements. Novel traits are not de novo constructions that depend on a series of genetic mutations. Third, phenotypic plasticity can facilitate evolution by the immediate accommodation and exaggeration of change. It should no longer be regarded as a source of noise in a system governed by genes, or as a 'merely environmental' phenomenon without evolutionary importance.[68]

Gilbert Gottlieb describes a similar path to evolutionary modification, one that begins with "a novel behavioral shift" that "encourages new environmental relationships."[69] The shift is initiated when animals

> live differently from their forebears. Living differently, especially living in a different place, will subject the animals to new stresses, strains, and adaptations that will eventually alter their anatomy and physiology (without necessarily altering the genetic constitution of the changing population). The new situation will call forth previously untapped resources for anatomical and physiological change that are part of each species' already existing developmental adaptability.[70]

For more than a century, mutants have lorded over the evolutionary domain. Clearly, they will not surrender their preeminent position without a fight. In the meantime, we should be wary of our reflexive tendency to appeal to the mutant when we encounter novel forms. Once again, consider the extreme case of dicephalus. A mutation is not required to produce this condition. Rather, as we have seen, dicephalus—*even* dicephalus—can arise with surprising ease through the alteration of an epigenetic process that, in the course of evolving the capacity to produce one head, incidentally evolved the capacity to produce two. We might even go so far as to say that dicephalus comes *naturally*.

THREE

Do the Locomotion
How We Learn to Move Our Bodies

We didn't lie to you, folks. We told you we had living,
breathing, monstrosities. You'll laugh at them,
shudder at them, and yet, but for the accident
of birth, you might be even as they are.
They did not ask to be brought into the world,
but into the world they came.
Tod Browning's Freaks (1932)

I could not stop looking for the rest of him. I saw a face—the handsome face of a movie actor—and a chest and arms. But I saw no legs. His tuxedo-clad torso stopped abruptly at his waist. As he rested on a table, motionless, I imagined him as a Thanksgiving meal, the main course served on a platter, awaiting the carving knife. But then, with gloved hands, he lifted himself up and began to walk, the empty space between figure and ground confounding my senses. Although moviemakers today are capable of producing astounding special effects, this was no special effect. This character was acting in a movie, *Freaks,* that was made in 1932.

Considered by many a masterpiece of the grotesque, *Freaks* provides a disturbing glimpse into a sideshow world that has virtually disappeared. And Johnny Eck, billed in *Freaks* as "the half boy," was part of that world.[1] For me, this man's every moment on screen is a revelation, each scene challenging my expectations as to how the human form ought to move. In *Freaks,* Eck's movements are so fluid, so *natural*—and yet his condition appears the opposite of natural. Like a duck with oversized feet, he bounds over a log, clambers up steps, and balances with impossible grace on a bedpost. He seems perfectly adapted to his condition—and in a sense, he is.

The distinctive world and astounding characters that *Freaks* depicts eclipse the movie's plot. Besides Johnny Eck, there are two "armless wonders"; Prince Randian, the human torso; a "she-man"; several "pinheads"; midgets; conjoined twins; a bearded lady; and a "living skeleton." Some are central to the story. Others merely show up long enough to display some remarkable skill—for example, an armless woman using her feet to eat with a fork, or Prince Randian famously using only his mouth to light a cigarette. These are not actors so much as acts, sometimes awkwardly inserted into the movie, designed to horrify and amaze, just as they did when performing for live audiences in sideshows and other venues.

At one level, *Freaks* is a melodrama that focuses on the relationship between Hans and Frieda, two midgets engaged to be married. When an attractive, nonfreakish trapeze artist seduces Hans away from Frieda so as to marry him, poison him, and inherit his fortune, she starts down a path that leads to her horrible disfigurement at the hands of the avenging freaks. But at another level—the one I found most fascinating—*Freaks* is about adaptation and the relativity of normalcy. As we watch the movie, we see that the inhabitants of this sideshow society exhibit the same range of decency, aptitude, and attractiveness that exists in society at large.

This isn't to say that the producers of *Freaks* were particularly respectful toward their actors. On the contrary, a message that was originally displayed to theater audiences before each screening revealed their true attitudes. It included this sentence: "Never again will such a story be filmed, as modern science and teratology are rapidly eliminating such blunders of nature from the world."

Although we have made great strides in preventing some developmental anomalies, we are also creating new conditions that make such anomalies more likely. Widespread industrial dumping of biologically active chemicals and global climate change are among the many environmental disturbances that can dramatically alter development. The accident at the Chernobyl nuclear facility in 1986 has left a tragic legacy that includes the birth of thousands of malformed children. Clearly, we are far from achieving the blunderless world envisioned by the producers of *Freaks*.

As for the "blunder" that was Johnny Eck, he was born without functioning legs but grew naturally into a hand-walker, exhibiting graceful movements with no sign of handicap or exertion. Recently, a different kind of hand walking was discovered within a single family living in an isolated village in southern Turkey. In this family of nineteen children, five siblings—all with fully formed arms and legs—crawl on hands and feet while bent over at the waist.[2]

The scientists studying these hand-walkers believe them to be mutant throwbacks to an ancestral walking style—"missing links" in our locomotor history. They argue that because hand walking emerged in each of these five siblings, they must possess a common "memory" or "program" that compels them to walk as our ancestors did. Furthermore, they argue that like dinosaur bones buried and forgotten in the desert, this memory was also buried—but within each of us. Now, we are to believe, this ancient memory has suddenly and remarkably resurfaced in a Turkish village. Thus, we see how the unrelenting allure of the missing link can inspire the search for a genetic mutation with the power to turn back time.

This theory is doubtful. Contrary to what one might expect of a locomotor throwback, the actual gaits of these individuals do not resemble those of our nonhuman primate cousins.[3] Rather, these individuals' distinctive locomotor style appears to solve a particular problem—a problem that begins with a severe malformation of the cerebellum, a part of the brain that contributes critically to balance and coordination. Coping with mental retardation and lacking parental encouragement to walk upright, these siblings move about the world the best way they know how. Indeed, the locomotor pattern that they have adopted is similar to that exhibited by many normal human infants—variously referred to as a "bear crawl" or "running on all fours"—before the transition to upright walking.[4]

What lessons should we in fact learn from the Turkish siblings? Here we reach a fork in the road. Follow one path and we accept at face value the notion—based on the most superficial of similarities—that the locomotor pattern of our four-legged ancestors has been reanimated within the bodies of a few severely disabled Turkish children. We will choose the other path, which is more firmly grounded in biological reality. Several companions will come along on this trip. They include, among others, Johnny Eck and that two-legged goat.

WALKING THE WALK

The list of key evolutionary developments that set humans apart from our biological cousins is short and widely known: an opposable thumb, our lack of fur, a relatively large brain, a unique capacity for language—and the ability to walk upright on two legs, that is, *bipedally*. Indeed, it is contended that bipedalism was a major turning point in human history because it spurred at least two further significant developments: the freeing of our hands for other uses (including gestural communication—the precursor to vocal communication—and human language) and the creation of a stable platform for an enlarged brain. The complete details of this narrative have yet to be fully resolved. But who can deny that human nature, whatever it is, rests upon bipedalism as surely as our large head rests upon our square, upright shoulders?

Having elevated our walking posture to such lofty heights, how are we to respond to news of a deformed goat that accomplished in its brief lifetime that which our ancestors presumably required scores of generations to accomplish? Bipedalism is not something that we humans stumbled into. Walking upright requires elegant coordination of bone, tendon, muscle, and behavior—coordination that cannot, we would think, arise spontaneously in an individual. So, our goat may have been able to walk upright—but that does not necessarily mean that it did so with efficiency and grace.

Still, our bipedal goat rears up to challenge our assumptions. It is famous among scientists because it did not merely walk upright. It developed the capacity to walk upright. As the goat developed without forelegs, its spine curved, its bones reacted, its leg muscles thickened, its tendons adjusted. With each moment in developmental time, anatomy shaped behavior and behavior shaped anatomy, the give-and-take like a conversation between longtime friends. So, like us, this goat did not stumble into bipedalism. *It was built for bipedalism.*

So too is the dog named Faith, reported from Oklahoma City in 2003 to have been born with severe foreleg deformities but able to walk or skip along on its two hindlegs.[5] There are also accounts of a wild baboon with undeveloped forelegs that learned to walk and jump in a bipedal fashion.[6] These and other observations testify to the "self-righting properties"[7] of developing systems and lend credence to the suggestion that "the evolution of bipedalism in humans might not be as difficult or as large an evolutionary step as some anthropologists have believed."[8]

Faith, the two-legged dog.

But even if we accept the reality of the two-legged goat or the two-legged dog, we are still left with the mystery of creatures belonging to four-legged species accomplishing, by hook or by crook, such a radical reorganization of body and behavior. How is it that a head that would normally be suspended at the end of a horizontal body could so easily become perched, and then balanced, at the top of a vertical one? This is not an insignificant rearrangement of parts.

Imagine lying in bed, balancing on your fingertips a plate of marbles. Then imagine moving to a sitting and then standing position without letting any of the marbles fall to the floor. Focus on the fine adjustments of wrist, arm, legs, and torso that maintain the plate's balance during these transitions. Concentrate on the segments of your body as they effortlessly adjust, correlated parts in motion, maintaining sublime balance as your center of gravity travels vertically during your ascent.

Now imagine our goat's remarkable path to bipedalism: The path begins as anatomical, and then behavioral, adjustments to the absence of forelegs. The path ends with a "malformed" animal nonetheless displaying the hallmarks of efficient and effective *design*.

Yes, design. But it is not design in the way we normally think of it and certainly not in the way creationists wield that term. The design of the bipedal goat was only *apparent* in the sense that it developed without oversight. But this judgment of apparent design does not make the goat's achievement any less astounding. Indeed, this simple, malformed goat has presented something of a challenge even to committed evolutionists.

Let us consider the options: One might imagine that the adaptability of the goat to two-leggedness reflects an evolutionary process *that has seen this problem before.* Perhaps buried within the animal—perhaps buried within every animal—are solutions to past problems that are suppressed or forgotten until needed again. Born without forelegs? Not a problem: simply activate the developmental program that adjusts to life without forelegs.

The notion that we can recover memories of long-forgotten developmental programs—invoked to explain the Turkish hand-walkers—may seem reasonable. However, in considering Johnny Eck, this "recovered memory" explanation doesn't even get off the ground: We would never suggest that Eck was a *throwback*—or that his walking style was the *instinctive* solution to an ancient problem faced by ancestors born without a functioning bottom half. Moreover, because bipedalism evolved relatively recently in mammals, we can safely assume that the goat's bipedalism had nothing to do with any putative bipedal habits among the goat's ancestors.

At this point in our deliberations, we might choose to adopt two separate explanations for the two sets of walkers: recovered memory for the Turkish siblings, and developmental adjustment for Johnny Eck and the two-legged goat. But in science we strive for *parsimony*, meaning that we look for the simplest answer that can account for as many facts as possible. In this case, the most parsimonious answer is clear.

Only one perspective ties together the Turkish hand-walkers, Johnny Eck, and the two-legged goat: Humans and other animals have evolved bodies and nervous systems that are rich in possible solutions to unforeseen problems. But these solutions do not lie dormant in the form of a static memory or prescribed program. They emerge dynamically through an unscripted process that reflects the inherent flexibility of complex systems. This is a developmental process, of responses nested within responses, of give and take. And when, as so often happens, we are surprised by what unfolds, we begin to appreciate the fact that *there is always more developmental potential than we know.*

Note that the goal of a developmental process is unforeseen. The developing animal cannot glimpse its destination over the horizon because its destination does not yet exist. What does exist is the developmental process itself. Each step along the way involves many choices, and with each choice, some opportunities are gained and others are lost. These opportunities represent our developmental potential, which fuels the evolutionary engine. Critically, regardless of the organism's assumed final form, each generation must make these choices anew. Perhaps it will make the same choices, perhaps not. But this much is clear: It is the full array of choices made—not merely the final form—that defines the fully grown adult. Thus, if children are to resemble their parents, their choices at each developmental step must be similar to those made by their parents.

Animals are not mere collections of traits, with each trait corresponding to a gene that evolved to produce it. They are instead the product of development that arises from an interlocking matrix of molecular mechanisms—including but not confined to DNA—which in turn form a web of interactions that bears a greater resemblance to conversational crosstalk at a cocktail party than to a prepared lecture delivered by a college professor.

What distinguishes biological entities from mere physical ones is a means of inheritance—again, including but not confined to DNA[9]—and a process whereby some animals survive or reproduce more successfully than others. But the key issue for us right now is straightforward: Animals do not simply show up. They do not pop into existence. They develop—and as long as animals develop, development will evolve.

I must be clear: When biologists highlight the bipedal exploits of, for example, a two-legged goat, they are not necessarily implying that such dramatically altered individuals provide the impetus for the evolution of a new species.[10] That is not the key lesson to be learned. Rather, as Pere Alberch explained, such examples

> illustrate how the nervous, skeletal and muscular systems interact with each other to accommodate perturbations in any of the component elements. In fact, I believe that this regulatory property is a general principle of development. The evolutionary implications of these phenomena have been

greatly underestimated, with most of the emphasis being placed on the genetic basis of morphological change. However, the regulatory capacities of an epigenetic system imply that any intrinsic change will trigger a sequence of regulatory changes to automatically generate an integrated phenotype.[11]

Thus, whatever caused the goat to lose its two forelegs—whether genetic mutation or environmental perturbation—the developing system was able to accommodate the change to produce a mobile adult. This *phenotypic accommodation*[12] may not presage the evolutionary success of two-legged goats *per se*, but it clearly reveals fundamental processes at work in normal animals as locomotor skills develop. It also reveals "the latent evolutionary potential of developmental systems," providing a sense of the internal constraints and biases that can shape evolutionary change.[13]

The choice that confronts us is a simple one. We can continue to carve animals up into discrete chunks and tell stories about the origins of each trait. How did the leopard get its spots? Rudyard Kipling's Just-So Story about an Ethiopian applying the spots with his fingertips is, to my mind, no more fanciful than the suggestion that a single family in southern Turkey has mutated into a quadrupedal human ancestor.

Alternatively, we can move beyond storytelling to take development seriously. When we do, creatures that once seemed so distinctively odd and incomprehensible suddenly become integrated into the larger community of organisms. It is not enough that Johnny Eck and the two-legged goat capture our attention. When properly considered, they also turn our attention away from those popular, simple, but misguided notions of how animals work and how they are constructed. Ultimately, the lessons learned about locomotion in anomalous individuals will help us to appreciate the distinctive locomotor patterns of individual species. Let us start by examining the principles that govern locomotion itself.

RULES OF MOTION

We do get around. We crawl, walk, skip, and run. We swim, dive, bike, climb, drive, ski, and skate. If any of these activities proves dangerous, we limp, hobble on crutches, or roll around in wheelchairs.

As remarkable as this locomotor diversity is, the subtler details of locomotion are just as captivating. I was reminded of this several years ago, soon after writing a book about the development of instinctive behavior. During a walk with my dog, a black-and-white Border collie named Katy, I became fixated on the patterns produced by her well-muscled legs, her gait shifting as she sped up and slowed down. I asked myself: How did Katy know to match a particular gait with a particular speed? To take a more familiar example, how does a horse know when to switch from a walk to a trot or from a trot to a gallop? Is this knowledge hardwired? Instinctive? I then wondered: How could I have neglected to address this topic in my book?

I tried to console myself by noting that few readers would expect to learn about the intricacies of walking and running in a book about instinct. Still, I knew that the patterning of limb movements as we walk and run provides useful insight into the rules that guide the development of behavior.

The day after my walk with Katy, I was standing at the front window of my house, reconstructing the internal dialogue of the day before. Then, out of the corner of my eye, a movement on the sidewalk caught my attention. Limping down the sidewalk, tethered to its owner, was a three-legged dog! Although its movements were not particularly graceful, this dog had devised a hobbling gait that compensated for its missing foreleg.

Watching Katy walking on one day, and the three-legged dog limping by on the next, I recognized, as I never had before, the common thread that ties all of us together. Katy could not have been born with the knowledge of how to switch gaits at different speeds. The dog passing by my window could not have been born with the knowledge of how to limp on three legs should it happen to lose one of them. Nor could Johnny Eck or the two-legged goat have been born with the knowledge to move about the world as they did.

As a dog walks around the house, climbs stairs, or chases rabbits, its *footfall patterns*—that is, the order and timing of leg movements—reflect the need to maintain bodily stability without wasting energy. To satisfy the demands of physics and efficiency, land mammals use a diversity of gaits. These gaits represent universal organizational principles of behavior. Understand the rules and the behavior follows naturally.

First, imagine a hypothetical one-legged animal. Like a pogo stick, such an animal could only hop. But add a second limb and things get a bit more interesting. Although hopping remains an option, walking and running are now options as well.

When you and I walk, we are like inverted pendulums, swinging from stride to stride.[14] Our feet swing forward with each step, establishing contact with the ground and providing the foundation for the next one. Enhancing this pendulum effect is our straight-legged style of walking, a distinctly human feature that takes pressure off of our leg muscles. Try walking with bent legs and you can immediately sense the strain. You can also see why chimpanzees, which on those relatively rare occasions when they walk upright do so with bent legs, prefer to move around on all fours. Already we see that walking, whatever its form, is intimately tied to the efficient use of energy.

I directly experienced the truth of this last statement when a tall, fast-walking friend and I spent a weekend gambling in Las Vegas. We were walking down the strip, and suddenly he was ten feet ahead of me. I quickened my step to catch up, but to no avail. With each stride, the gap between us grew. I broke into a slow run to catch up and then transitioned back to a walk. But then he was ahead of me again. And so it went: walking, running, walking, running, all the way down the strip. He was oblivious. I was exhausted.

It seems that we all have a range of walking speeds within which we feel most comfortable. Too slow and we feel awkward. Too fast and we feel like we might topple over at any time. Indeed, as we speed up, the effort increasingly seems out of balance, as if we are somehow putting in more than we are getting back. Then comes that point—and we all know it when it happens—where we feel the urge to break into a run. For adults that point usually occurs around five miles per hour.

The feeling of instability as we raise our walking speed is linked to our decision to transition from walking to running. Typically, as we walk, the gravitational force provided by our planet interacts with the length of our legs to limit our maximum walking speed. The rules are simple: The longer our legs or the stronger the force of gravity, the faster we can comfortably walk. This is why small children must run to keep up with their parents. It is also why I was having such trouble that day keeping up with my friend in Las Vegas.

Once we are off and running, new rules apply. Gone is the pendulum swinging from side to side, and in its place is a spring bouncing around with a weighty object on top. Replacing the straight legs of walking with the bent legs of running, we bounce along with each step. With each bounce, tendons and ligaments in our legs are stretched and the arch in our foot is compressed. The "spring" is compressed and energy is stored for the next running stride. Then, the spring is released: We take off, aided by the release of stored energy. The next foot lands, and the process of spring compression and release repeats. Thus, at its core, human running takes advantage of similar physical processes as hopping in kangaroos and rabbits, except that our feet land in alternation rather than in unison.

Although no human can outsprint a gazelle or a cheetah, we are remarkably able endurance runners.[15] Combining high running speeds with sustained endurance over many hours, we humans compare favorably with many other species, such as African hunting dogs and wolves, that are built for running long distances across open habitats. What makes this human capability possible is a suite of anatomical and physiological adjustments that we possess uniquely among primates, including springs in our leg and foot (including the Achilles tendon), an arched foot (except for those of us with flat feet), and enlarged gluteus maximus. We are also among a rare set of species that uses a single running gait across a continuous range of speeds without any added energy cost. These and other facts support the notion that the ability to run long distances may have been a critically important factor in human evolution.

When we add two more legs to the locomotor mix, things get more complicated. With more legs come more combinations. For example, the standard walking pattern in horses comprises movements in the following sequential order: left foreleg, right hindleg, right fore, left hind. But when the horse starts to trot, diagonally opposite pairs of legs move in unison: left fore with right hind, right fore with left hind. With galloping, the two fore limbs hit the ground in close succession, followed by the two hind limbs. If you listened to the sound made by each foot as it strikes the ground, you would hear four evenly spaced beats in the walk (tick, tick, tick, tick), two evenly spaced beats in the trot (tick, tick), and four unevenly spaced beats in the gallop (tick-tick, tick-tick).

How do we explain this intricate relationship between footfall pattern and speed of locomotion? An answer to this question came in 1981 when investigators successfully trained ponies to walk, trot, and gallop on a treadmill at various speeds while their energy use was measured.[16] This was no easy task for researcher or pony, but the results were well worth the effort.

The researchers found that each gait—walk, trot, and gallop—was associated with a narrow range of speeds within which energy consumption was minimized. For each gait, maintaining speeds above or below its associated range demanded higher energy costs. The researchers proved the validity of this rather artificial treadmill experiment when they observed the same ponies in an open arena. Consistent with their treadmill behavior, the ponies precisely matched each gait to a narrow range of speeds.

In addition, as the ponies rambled freely about the arena, increasing and decreasing their speed, the investigators noticed that some speed zones were never occupied. The ponies simply skipped over these forbidden zones as they transitioned from walking to trotting, and from trotting to galloping. Just as I had switched abruptly from pendular walking to springy running as I chased my friend in Las Vegas, the ponies switched from the four-beat walk to the two-beat trot to the four-beat gallop without, well, missing a beat.

One more point: Recall that limb length helps to determine the speed at which we transition from walking to running. As it turns out, this principle also applies to different animals. Thus, large animals, such as horses, make the transition from walking to trotting at a faster speed than do smaller, shorter-legged animals, such as dogs. Indeed, the physical principles are so well understood that we can predict that a horse whose limbs are nine times longer than a dog's will go from walking to trotting at the square root of nine, or three, times the speed at which the dog will.

Now, returning to my earlier question, how does an animal know how and when to switch gaits? Conversely, how does an animal know which speeds lay within the no-man's-land between walking and trotting, and between trotting and galloping? Katy knows it. Ponies know it. In fact, every biped and quadruped knows it. It is untaught *and* universal.

There are of course other concepts that might be offered to explain the pony's elegant, fluid, and seemingly spontaneous transitions between walking, trotting, and galloping: *innate, genetically programmed, hardwired, predetermined, scripted, instinctive.* But such concepts provide only the illusion of explanation. Given that gait transitions occur at higher speeds in longer-legged species, in longer-legged individuals within a species, and in all of us as our legs grow, it is implausible to imagine that such highly detailed knowledge about the matching of gaits and speeds is preprogrammed into the brain of a newborn. Leaving aside how such information could be transmitted or implanted, the sheer quantity of information needed is staggering. At a minimum, the newborn would have to foresee the length of its legs *at each stage of development.* To be able to do that, it would have to know how fast it would grow and, therefore, how well it would be fed. Clearly, no animal possesses such foresight. DNA is not clairvoyant.

In fact, the developing fetus cannot even foresee the planet on which it will be born or will live out its life. At least as regards locomotion, this limited vision appears not to matter. Humans can walk and run just about anywhere, even on the moon. Only the details change: On the moon, with its weaker gravitational force, transitions between walking and running occur at lower speeds. But they still occur.

Leaving aside explicit foreknowledge of particular gaits and speeds, perhaps there exist general locomotor rules that are instinctive. But this is also unlikely, whether the rules are *If you are a goat born without forelimbs, stand upright and walk like a human,* or *If you are a human born without functioning legs, walk on your hands,* or *If you are alive, walk, trot, gallop, or otherwise move so that stability is maintained and energy is used as efficiently as possible.* Such rules are just another way of saying that each of us, at each stage of life, discovers what works best for us—and that is a far cry from a preprogrammed, genetically controlled instinct.

The bottom line is this: The ability to locomote—to walk, trot, run, hop, swim, and fly—arises through moment-to-moment interactions among hard and soft parts, bone and brain, ligament and learning, anatomy and behavior. These fundamental behaviors may often appear to be instinctive, inborn, innate, hardwired, or predetermined, but in fact they are *discovered*—by each of us, individually, in real time—through a developmental dialogue among all our parts. Locomotor skills, even those that are untaught and universal, arise from the complex interaction of these parts. Just as we learn how to sprawl in the heat and curl up in the cold, and just as males and females discover the pros and cons of standing while urinating, each of us discovers at each point in our lives how our body moves and interacts with the surrounding environment. Out of this continual interaction, we acquire stability and efficiency in our movements. If we lose a limb, walk on stilts, travel to Mars, or experience a growth spurt, the discovery process begins anew.

The rules are simple, untaught, and universal, but the range of possibilities they engender, limited only by the realities of physics and chemistry and by our evolutionary history, is far broader than we often recognize. There is no script. Flexibility is key.

FORM AND FUNCTION

Scientists long recognized an intimate relationship between morphology—the shape and size of body and limbs—and loco-motor behavior. It is easy to appreciate how the aerodynamic design of wings and the hydrodynamic design of fins are vital to life in air and water, respectively. Similarly, the over-developed hindquarters of kangaroos, the elongated arms of gibbons, and the webbed feet of beavers and water dogs all reflect the distinc-tive locomotor lifestyles of the animals that possess them.

But how do we account for the match that we see in so many different species between morphology and behavior? Does one influence the development of the other, or do they arise independently? For example, one might imagine that kangaroos hop because they have overdeveloped hindquarters *and* because they have an instinct to hop. After all, kangaroos are hoppers, just as cheetahs are sprinters and humans are upright walkers. According to this view, the exquisite match between morphology and behavior reflects the action of an evolutionary process that produced each feature independently. But this view seems most plausible when we restrict our focus to the adult, "normal" members of a species.

Freaks eloquently testify to the co-dependent relationship between morphology and behavior. But we should not forget why Johnny Eck and the two-legged goat had such a significant advantage over similarly challenged individuals who are injured or maimed as adults: They grew up with their bodies and adjusted accordingly—so much so, that it becomes difficult to see them as disabled. Indeed, if we did not know what a typical human or goat looked like, we might hardly miss their missing parts.

But the deeper truth is this: The extraordinary locomotor abilities of Johnny Eck and the two-legged goat are no more extraordinary than those that develop in every other animal on our planet.

Thus, behavioral adjustments to an atypically formed body parallel the adjustments that all of us make as our typical bodies are transformed throughout development. These transformations can easily escape our attention: Like the movements of a glacier, development—especially human development—is slow, and changes are subtle. So it is no wonder that we know much less about locomotion in developing animals than in adults. Still, based on the developmental studies that have been conducted so far, we can say that the processes that characterize locomotor development are quite remarkable.

We can begin to appreciate locomotor development by considering the factors that shape human adult locomotion. For example, at a scientific conference ten years ago in Jackson, Wyoming, a group of us decided to explore Yellowstone National Park on snowmobiles. One of us drove twenty yards off the path and got stuck. I tried to walk toward the stranded machine to help out, but with each step, my legs sunk deeper in the snow. I fell forward, righted myself, and continued, which was exhausting and futile. Then, before I knew what I was doing, I was on my hands and knees, *crawling* toward the snowmobile. Under these particular conditions, crawling was more comfortable and effective than walking.

The forces that shaped my regression to crawling that day are similar to those that normally shape progressive locomotor changes in developing infants. All of us—infant and adult— discover through trial and error that some movements are more effective and efficient than others, depending on the terrain, our task, and even our shoes and clothes. With infants, their rapidly changing bodies complicate matters further, thus accounting for their particularly diverse array of locomotor patterns—they belly crawl, bum shuffle, knee walk, cheek crawl, log roll, goose step, and bunny hop. The same infant will even shift locomotor patterns from one moment to the next. In short, what characterizes infant locomotion is variability and idiosyncrasy.[17]

On the path to upright walking, some infants crawl on hands and knees—and some "bear crawl" on hands and feet, similar to the Turkish hand-walkers. There is not anything special about the bear crawl: It no more reflects locomotion in a four-legged ancestor than the belly crawl reflects a slithering ancestral past.

Every day, the typical infant that is learning to walk takes more than 13,000 steps—the length of nearly forty football fields—falling ninety times in the process.[18] Through such extensive daily experience, human infants learn how their bodies function in their terrestrial environment. As a result—and even as our infantile, then adult bodies shift in size, shape, strength, and structure—most of us arrive at similar solutions to our loco-motor problems. This idea can be exported to other species—including those with highly unusual locomotor patterns—to explain how they, too, adjust to the profound changes in body size and shape that characterize early development.

To investigate the developmental links between locomotor form and function, some species are more useful than others. Because time is precious, researchers wisely choose animals that make the transition from infant to adult in weeks rather than years. Rodents, such as rats, mice, and gerbils, are born after just a few weeks of gestation and are weaned several weeks after they are born. The speed of their transformation—from naked, deaf, and blind to furry, perceptually engaged, and independent—is astounding. They provide a wonderfully succinct opportunity to examine the moment-to-moment links between morphology and behavior throughout development. Moreover, the diversity of this group of animals gives us a chance to detect differences in the way each type of rodent develops.

Take, for example, voles, gerbils, dormice, and jerboas.[19] As adults, these rodents differ where it counts, at least for our purposes. Voles have equally sized forelegs and hindlegs attached to a squat, cylindrical torso; gerbils have a short, arched torso, and hind legs that are about twice the length of their forelegs; dormice have a wide, squat trunk and, like gerbils, hind legs that are twice the length of their forelegs; and jerboas—the oddest of the bunch—have a trunk like dormice but are equipped with outsized hind legs that grow to four times the length of their forelegs.

Typical postures and body shapes of four rodent species over the first three postnatal weeks (and beyond for the jerboa). Note the similarities at birth and the diverging characteristics thereafter.

The adult movements of these animals are equally diverse: Voles move primarily by *trotting*, each foreleg moving in synchrony with the diagonal hind leg; gerbils move primarily by *bounding*, the two hind legs pushing off the ground together, followed by the two forelegs landing together; dormice trot but also *gallop*, similar to the bound except that the two hindlegs (and forelegs) do not hit the ground simultaneously but rather in quick succession; and jerboas walk *bipedally* (though, like kangaroos, they also *hop*).

Based on just this cursory description of these four rodents, the links between morphology and locomotor behavior begin to reveal themselves. For example, the relative length of the hindlegs and the shape of the torso constrain the kinds of locomotor gaits that will be possible or preferred. Trotting favors forelegs and hindlegs of similar length, bounding is aided by an arched body centered over enlarged hindlegs, and bipedal walking is possible when an animal is able to rise up on a steady foundation provided in part by oversized hindlegs. But if we truly wish to understand the roots of these striking differences in adult morphology and behavior, we must examine the earliest stages of development.

At birth, all four species are almost indistinguishable in form—forelegs and hindlegs are of equal length and their infantile bodies lie flat against the floor. As newborns, they also all move using a similar maneuver. In this maneuver, called *punting*, the pup moves one forepaw over the other and, because the hind legs do not move, the pup simply rotates around its rump. The resulting circular pattern of movement may not be effective for linear travel, but it is suitable for the movements required of pups within the nest.

As pups gain the leg strength needed to lift their torsos off the surface, the next locomotor pattern to emerge in all four species is walking. With increased strength and stability comes trotting, which for voles is the final gait in their developmental sequence. For gerbils, trotting is followed by bounding, a locomotor milestone that is accompanied by elongation of the hindlegs and arching of the back. For dormice, trotting is followed by galloping; for jerboas, bipedal-walking comes next; and for dormice and jerboas, as with gerbils, the emergence of their specialized modes of locomotion mirrors the lengthening of hind legs. This lengthening is particularly dramatic in the jerboa in which the hind legs are the same size as the forelegs at birth, but four times longer only seven weeks later.

So there are clear developmental patterns at play: Starting at birth when morphology and behavior are highly similar, the members of each species follow a shared sequence of events: from punting to walking to trotting—to bounding (gerbil), galloping (dormouse), or bipedal walking (jerboa). Each developmental transition typically depends upon the emergence of a suite of bodily characteristics.

For example, to make the transition from walking to bounding, a young gerbil needs more than just strong hind limbs.

When it bounds into the air, its forelimbs also must be strong enough to absorb the impact of landing. If the forelimbs are not sufficiently strong, they will collapse and the gerbil's head will smash into the ground. To avoid such mishaps—or perhaps in response to them—gerbils briefly employ an intermediate gait, called the half-bound, in which the two forelimbs hit the ground in succession rather than simultaneously. This slight modification appears to do the trick during this brief period of vulnerability. Then, within a few days when the forelegs have increased in strength, the bound replaces the half-bound.

The sense that locomotor development is a process of personal discovery is heightened when we consider the jerboa's path from birth to adulthood. As already mentioned, the hindlegs grow rapidly during early development and the pup must cope with these rapid changes. Initially, as the pup creeps around using only its forelegs, it resembles the other species that we have discussed. But within weeks, as the hindlegs lengthen but are not yet able to support the body, they simply hang to the animal's side while the forelegs do all the work. If we did not know better, we might think that the young jerboa, with its freakishly elongated and ineffectual hindlegs, is the victim of a disabling condition. But this condition soon passes as the hind legs are brought under the body and the pup transitions to quadrupedal walking. The pup is on its way to upright walking, but it is not there yet.

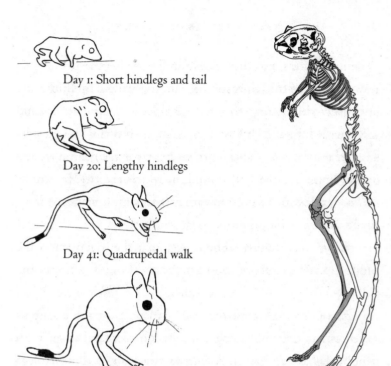

Day 1: Short hindlegs and tail

Day 20: Lengthy hindlegs

Day 41: Quadrupedal walk

Day 50: Horzontal posture of the trunk

**Typical postures of the jerboa from birth through fifty days of age (left).
The skeleton of an adult jerboa (right).**

It is at this stage, at about 40 days of age, that a curious thing happens. With hind legs roughly three times the length of its forelegs, the pup is at an awkward stage. As it tries to walk on all fours, its oversized hind legs interfere with walking, not unlike the difficulty we face when we walk with flippers on our feet. Initially, the tips of the flippers catch on the ground with each step, but we soon learn the necessary adjustments, lifting and angling our legs and ankles. The result may look goofy and feel awkward, but it works.

Similarly, during their transition from creeping to walking, jerboas learn to fold their hindlegs up toward the body, thereby effectively shortening the legs and making quadrupedal walking possible. In a few more days, as hindleg bones and muscles strengthen, the pup is able to lift its rump further off the ground and walk more effectively. Soon, the pup is able to sit securely on its hindquarters, providing a necessary foundation for its final transition to upright, bipedal walking.

This developmental progression in jerboas illustrates the bidirectional causal relationship between morphology and behavior—between form and function. To think instead that this relationship is unidirectional is to misread the nature of development, like imagining a unidirectional causal arrow from gene to behavior. Rather, form and function continually feed off of each other through developmental time. Limbs lengthen and bones harden. New gaits emerge. Muscles strengthen. Locomotion becomes more efficient. Endurance increases. The animal explores new worlds.

If development tells a story, then it is like those ghost stories that children tell around a campfire, with each youngster adding a snippet, building on what has come before. There is no single narrator and no internal dialogue. There is what just happened, what is happening now, and what comes next.

———

WHEN WE SAY that the skull is a hard part and that the behavior that is controlled by the brain within that skull is a soft part, we are commenting as much on the ease of molding them as on their material composition. Indeed, depending on the circumstances, even bone can be a soft part: As we saw with those human cultures that encouraged manipulation of skulls or feet, bone is easily moldable, but only early in development when growth is rapid. Similarly, behavior is not equally pliable throughout our lifetimes. As with bone, *behavior is most easily shaped during early development.*

If we wish to fully understand the forces that shape early locomotor behavior, it is not enough to simply observe developmental changes in bone and muscle. We must also decode the neural circuitry within the spinal cord and brain that generates movement. Our goal is to understand how these neural mechanisms detect incoming sensory information, process that information, and send outgoing signals to muscles to produce movement.

Motor and sensory signals to and from our legs are processed within the lower portion of our spinal cord; signals to and from our arms are processed within the upper portion. Within each segment of the spinal cord are clusters of neurons that, when activated together, cause individual muscles in our limbs to work. Indeed, we find a close relationship between the activity of these neurons and the activity of the muscles that they control. For example, in rabbits, neurons within the lower spinal cord that control both right and left hindlegs are activated together, thereby producing a hop. This is not terribly surprising—hopping legs should be controlled by "hopping" neurons.

To determine whether neurons within the spinal cord alone can produce these hopping rhythms, researchers cut the spinal cord of adult rabbits so as to isolate the neurons that control the hindlegs from the rest of the nervous system.[20] Although such animals are functionally paralyzed, certain drugs are able to directly activate the spinal cord neurons, thus stimulating movements of the otherwise paralyzed limbs. Surprisingly, when these paralyzed limbs are stimulated to move, they exhibit organized stepping movements that follow the pattern of walking, not hopping.

Why would the fundamental neural circuitry that controls the muscles in the hind legs of a hopping mammal exhibit a walking pattern? If hopping in rabbits were instinctive, as many would assume, would we not expect that animal's fundamental motor-control system—which lies within the spinal cord—to exhibit the hopping pattern? But this research instead suggests that the spinal cord readily expresses the walking pattern of locomotion and that the brain overrides the spinal cord to produce hopping. In a nutshell: The spinal cord "walks" and the brain "hops."

To explain this arrangement, we might wonder if the hop arrived late on the evolutionary scene, by which time the predominant locomotor pattern was firmly ingrained within the neural circuitry of the spinal cord. Such a scenario might explain the "walking" spinal cord and the need for the brain to take control to produce hopping. At its heart, this *style* of explanation invokes ancient, ingrained instincts that are impervious to modern-day experience.

Evolutionary psychologists promote similar stories about the historical roots of human behavior. These theorists argue that such traits as jealousy, fear, and even religiosity were established hundreds of thousands of years ago when our ancestors were hunters and gatherers. Moreover, they argue, attitudes and cultural norms may change, but such changes cannot suppress indefinitely our compulsion to act on these ancient, instinctive inclinations. But even when we examine something as tangible as a neural circuit for producing such basic behaviors as walking and hopping, these evolutionary stories—popular though they may be—fall flat.

We move closer to appreciating the rabbit's locomotor system when we examine the development of walking and hopping. Having reviewed this process now in several other small mammals, you can anticipate that quadrupedal walking, involving alternating movements of the hindlegs, predominates in the early lives of rabbits as well. For them, hopping replaces walking only by the age of three weeks. Do the young rabbit's alternating leg movements during walking somehow tune the neural circuitry within the spinal cord so that it preferentially produces an alternating pattern of neural activity? If so, the normal "walking pattern" of spinal cord activity would not be preprogrammed, but rather shaped by early experience. To test this possibility, researchers severed the spinal cords of two-day-old rabbits, once again disconnecting the neural circuitry controlling hindleg movements from the rest of the nervous system. (It should be noted that spinal cord injuries in young animals do not produce the drastic effects that they do in adults. This is because neural circuits within the spinal cord can function semi-autonomously for some time after birth. But once the brain establishes connections with the spinal cord, the spinal cord circuits lose their autonomy; now, severing the connection between spinal cord and brain produces a severe loss of function.)

What these researchers did next was clever, strange, and informative. When the rabbits were ten days old—eight days after their spinal cords were cut—they were exposed to one of several training regimens. They were secured in a harness and their two hind feet were strapped onto two motorized pedals that moved the limbs to simulate, for some rabbits, an alternating walking gait and, for others, a synchronized hopping gait. The rabbits received this training daily, in fifteen-minute sessions, until they were thirty days old. At that time, the rabbits were induced to kick their legs (by simply pinching the tail) as the activity of the leg muscles was monitored.

The results of this extraordinary experiment belie the notion that the spinal neural network is preprogrammed to express a walking pattern.[21] On the contrary, the hindleg movements at thirty days of age reflected perfectly the kind of training received. If the rabbits were trained with the alternating pattern, they exhibited a walking pattern of leg movements; if they were trained with the synchronous pattern, they exhibited a hopping pattern. Thus, early limb movements *tune* neural circuits.

————

WHEN I WROTE these words, I was still digesting a conversation that I had had with several colleagues earlier that day. We were discussing an issue that is germane to this chapter: How do human infants come to do the amazing things that they do so soon after birth? One colleague, a nativist at heart, tried to make sense of it this way: "Infants," he said, "don't have billions of years to develop." He was arguing that the enormity of time allotted to the evolutionary process is a luxury that individual animals cannot afford, and so they adjust to this time crunch by preprogramming critical behaviors and core knowledge into the newborn brain.[22]

But it does not follow that human infants must cram billions of years of organic evolution into a nine-month gestation period—any more than you or I need billions of years to learn to ride a bike or drive a car. Nor did Johnny Eck or the two-legged goat demand extra time to tackle the locomotor challenges posed by their oddly formed bodies; indeed, given that their bodies were unique and unpredictable, the notion that their locomotor skills might have been preprogrammed is preposterous. Rather, developing in lockstep with their bodies, novel skills emerged—efficient, graceful, and timely.

Still, the lesson to be drawn from this discussion does not concern only the emergence of novel capabilities such as those of Johnny Eck and the two-legged goat. The deeper lesson is that the developmental processes that gave rise to their capabilities are no different from those at work in each and every one of us. We are all extraordinary because of those processes.

Does the spinal cord develop such that, by the time of birth, it can support locomotion? Under ordinary circumstances, it does. But that structure is not prespecified to produce an evolutionarily ancient walking gait. As we saw with the young rabbits, spinal neural circuitry can be tuned to generate a walking or hopping pattern; and as we saw with jerboas, changes in limb length and strength can make old gaits obsolete and new gaits possible. All animals move about the world, but the fine details of how they move reflect a highly dynamic, epigenetic, and often idiosyncratic interplay between body, brain, and experience.

Try to imagine how Johnny Eck and the two-legged goat could have accomplished their locomotor feats in any other way. We can weave evolutionary "just-so stories" to make sense of them. But by doing so, we will find ourselves looking down upon Johnny Eck and the two-legged goat as little more than freaks. However, if we adopt the epigenetic perspective and acknowledge the interweaving interactions of form and function through developmental time, these anomalies are transformed before our eyes. We no longer stare at them with horror. Our shock is replaced by awe. And, perhaps more important, by a sense of kinship.

FOUR

LIFE AND LIMB
How Limbs are Made, Lost,
Replaced, and Transformed

*... we must not forget that what appears to-day
as a monster will to-morrow be the origin of a line
of special adaptations. The dachsund and the
bulldog are monsters. But the first reptiles with
rudimentary legs or fish species with
bulldog-heads were also monsters.*
RICHARD GOLDSCHMIDT (1933)[1]

*So you perceive it's really true,
when hands are lacking, toes will do.*
ANN LEAK THOMPSON, *"ARMLESS WONDER"*[2]

L ike branches on a tree, limbs reach outward. Whereas eyes,
ears, and nose specialize in the detection of distant signals,
limbs provide a more intimate link to the immediate environment.
With that link also comes the potential to move and manipulate;
feet on the ground, fingers on the keyboard, we sense and respond.

The loss of a limb, then, is a particularly profound trauma.
Indeed, to my mind, few things would be more horrific than

being a wounded soldier during the Civil War, awake and aware as a surgeon saws through your damaged leg. As the limb falls to the ground, a castaway in a sea of abandoned body parts, the nervous system seeks to adjust to sudden incompleteness. The adjustment is not easy.

The accidental loss of limbs accelerated rapidly during the nineteenth century, due in large part to the modernization of industry and warfare.[3] With this modernization of loss came a modernization of gain, as prosthetic arms and legs transformed useless stumps into functioning (or only more attractive) appendages. More recently, with steady improvements in armor and battlefield medicine, many injuries to U.S. soldiers that proved lethal in Vietnam have proved survivable in Iraq—producing a flood of limb injuries that, in turn, propel further advances in prosthetic technology.

For example, *myoelectric* prostheses allow muscles in the stump to electrically control the device. Still, even such devices lack direct sensory feedback from the prosthesis, so amputees must carefully watch their own movements and adjust accordingly. Therefore, neuroscientists and engineers are aiming to develop fully integrated *brain–machine interfaces* that let an injured individual's brain directly sense and control robotic prostheses.[4]

These interfaces are possible because of our progress in deciphering the computational codes that the brain uses to detect and control limb movements. Specifically, in the mature nervous system, every sensation from the fingertips (for example) produces discrete activation patterns in the brain, and discrete locales within the brain produce specific finger movements. Such *map-like* or *topographic relations* between muscle and brain are a general feature of nervous system function. Without such finely tuned relationships, we could not play a piano or type at a keyboard. And when a limb suddenly goes missing or the lines of communication between brain and limb are severed, the brain does not erase its representations of that limb (although those representations are modified). As a consequence, the brain is typically fooled into thinking that the limb remains attached, a phenomenon known as *phantom limb*.[5]

Losing an arm in an accident does not make one a freak. In fact, amputees were not among the stars of Victorian-era freak shows. But people *born* without arms were superstars. Like the two-legged goat, these *armless wonders* achieved fame for the way they adapted to their predicament:

> Sanders Nellis, a [P. T.] Barnum regular, cut out Valentines, wound his watch, loaded and fired a pistol, and shot a bow and arrow, all with his feet. Ann Leak Thompson was an exceptionally pious woman who worked religious symbols into her crocheting and embroidery.... And the famous Charles Tripp spent upward of fifty years doing carpentry, painting portraits, and practicing penmanship before appreciative audiences.[6]

It is as if their legs, feet, and toes were their prostheses—tailored for uses that most of us assume are the exclusive domain of arms, hands, and fingers.

Sanders Nellis, Ann Leak Thompson, Charles Tripp, and others were able to convert their idiosyncrasies into cash. But the financial benefits that they derived from their abilities were secondary to their basic need to survive. Indeed, long before the arrival of paying audiences, at least some armless individuals managed their daily lives. In 1528, for example, a German girl was reported to be living "entirely deprived of arms, the rest of the body very well formed. Her feet, by means of which she had thrived for many years, she used instead of hands, and marvelously well."[7]

So, where amputees typically use prostheses to mimic the actions of their missing arms, people born without arms develop compensatory capabilities in their remaining limbs. Moreover, whereas the traumatic loss of a limb in adulthood produces the sensation, even pain, of a phantom limb, such phenomena are distinctly absent in those born without limbs.[8]

Such qualitative distinctions between amputees and the congenitally limbless become even more apparent when we consider early attempts to merge their two worlds. In a recent documentary,[9] Gretchen Worden, former director of the Mütter Museum in Philadelphia—a museum famed for its collection of medical oddities and malformations—describes a photograph in its collection. The photograph depicts a young girl born with the tiniest of legs who walked, just like Johnny Eck, on her hands. "I love this little girl," says Worden. On the screen we see a photograph of the girl, her naked body hoisted in the air, her minuscule legs dangling uselessly. Discussing another photograph of this girl, Worden continues:

She has this wonderful straightforward, unaffected look at the camera. She's totally unselfconscious. And yet she has this extraordinary body with the deficiency of the lower limbs. You don't get the sense of any deformity [and] certainly not disability. She's perfectly able.

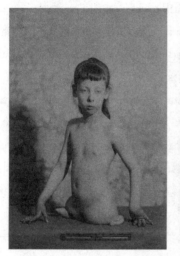

A young girl with amelia, a condition marked by undeveloped arms or legs. Johnny Eck had this same condition.

Then, changing gears, Worden recalls a third photograph of the same girl that was not displayed in her museum and is not shown in the documentary:

It shows her in a steel and leather prosthesis. It's sort of a bucket that goes around her hips. And then she has steel legs with little shoes on them. And all of a sudden she looks like a cripple.

It is ironic that a prosthesis would give the perception of crippling. But that is the inevitable consequence of any attempt to use a crude mechanical contraption to replace the adaptive power of development. Today, the use of prosthetics by infants and children with congenital limb abnormalities is on the rise.[10] But any such effort to modify the development of a child must be done carefully and thoughtfully.

For a child born with all her limbs, development follows a path that is typical of our species. But when a child is born with limbs missing, she traverses a distinct developmental path that entails novel brain organization and associated behavioral adjustments. Consequently, we would never expect an adult amputee to develop the extraordinary talents of an armless wonder. Such talents require all that development has to offer.

STAR MAPS

Fetal and infant brains exhibit an admirable adaptability to unique developmental conditions—whether those conditions are unique to individuals within a species (as with armless wonders) or are unique among species (as with jerboas). This adaptability is essential given the unpredictable size and conformation of our bodies. Predetermined brain function is neither feasible nor desired. Instead—reflecting a theme that is by now familiar—the functional organization of the nervous system reflects how it interacts during development with its particular sensory and motor systems.[11]

That individuals can exhibit compensatory responses to sensory loss might be familiar even to people who have not met an "armless wonder." Have you ever wondered whether the extraordinary musical ability of Stevie Wonder owes anything to sensory compensation resulting from his congenital blindness? There is now evidence that it does.[12] Moreover, sensory compensation is accompanied by brain reorganization. For example, in congenitally blind humans, the part of the cerebral cortex that, in sighted people, would process light arriving from the eyes is recruited to process tactile information arriving from the fingers.

This reorganization was demonstrated experimentally by disrupting the functioning of the "visual" part of the brain (by exposing it to powerful magnetic stimulation) as blind individuals used their fingers for Braille reading.[13] This stimulation distorted the tactile perceptions of these blind subjects, whereas similar manipulations of the brains of sighted individuals disrupted vision without affecting tactile perceptions. Thus, even after millions of years of processing visual information, the mammalian "visual" cortex remains open to inputs from other sensory systems.

This reorganization of the cerebral cortex also has been observed in short-tailed opossums that were experimentally blinded soon after birth.[14] When their adult brains were examined, sensory maps of the brain surface[15] showed extensive encroachment of the areas responding to sound and touch into the area that, in a sighted animal, would respond only to light. We might say that in the absence of visual stimulation, the "visual" cortex is *colonized* by other sensory systems.

·

A subterranean and visually challenged rodent, the naked mole-rat, is shown in close-up to reveal its oversized teeth used for digging (left). A cartoon (right) depicts how a naked mole-rat would look if its body parts were proportional to the amount of cortical tissue dedicated to them. Thus, for this species, the cortex processes a disproportionate amount of sensory information arriving from the head, particularly the incisors.

Incredibly, a parallel process of cortical colonization has been observed in the blind mole rat,[16] a species whose visual system has degenerated almost completely as an evolutionary response to life underground. Indeed, in this species, brain reorganization is particularly extensive, involving brain structures that feed into the cortex.[17]

The subterranean lifestyle of the blind mole rat reduced its demands for a high-functioning visual system. In its place, this animal—and other "tooth-digging" species, such as the naked mole-rat—evolved to rely on sensory information arriving from four front teeth.[18] A similar dynamic shaped the evolution of the star-nosed mole, a small insect-eating subterranean mammal that digs elaborate burrows in the wetlands of eastern North America. Like other species with reduced visual and auditory capabilities, these animals largely feel their way through life.

An adult star-nosed mole (left) emerges from its burrow, highlighting its impressively clawed forelimbs and the bizarre touch-sensitive appendages on its snout. The emerging star can be seen early in embryonic development (right). As with the blind mole rat's dedication of cortical tissue to its teeth, the star-nosed mole dedicates enormous cortical processing resources to star, as well as its clawed forelimbs (cartoon at bottom).

But what makes the star-nosed mole truly unique is the star-shaped organ—comprising twenty-two touch-sensitive appendages—that resembles a jester's hat jammed on the end of its snout. The appendages are controlled independently and move along the ground like fleshy fingers feeling their way in the dark. When potential food is detected, the star-nosed mole orients toward it so that the two appendages just above the mouth can take a "look" before the food is eaten.[19] All of this is remarkably rapid—less than a quarter-second from first touch to ingestion. Indeed, the star-nosed mole is among the fastest known mammalian eaters.[20]

The star-nosed mole's appendages are represented as a map—actually several maps—on the cortical surface. Moreover, as has been shown in other sensory systems—for example, with the whiskers of rats and other rodents—these maps are laid out in remarkably orderly fashion. Each of this mole's appendages is represented by a "stripe" on the cortical surface, and adjacent appendages are represented by adjacent stripes in the brain. It is as if the star itself has been dipped in paint and laid on top of the brain. The map is that true.

Even in the bizarre world of star-nosed moles, freaks can occur. For example, one mole was found with twelve append-ages on each side of its snout rather than the usual eleven.[21] When this mole's cortex was examined, its stripes also numbered eleven.

For all of the reasons discussed so far, it is clear that sensory organs on the periphery "instruct" the developing brain to pro-duce functional map-like representations.[22] But this instruction is not mindless. On the contrary, it appears that the final organi-zation of the cortex reflects the structure of the peripheral organ *and* the uses put to that organ early in development.[23] Thus, the cortical representations of the star's appendages near the mouth are larger than the others, reflecting their magnified importance as food's last stop on the way to the mouth.

We know from research in human adults that experience—such as playing a stringed instrument[24]—can modify cortical organization. Such experience is even more profound in early infancy, when the most fundamental relationships between peripheral structures and the brain—and among systems within the brain—are being established. When new sensory structures arise—as with the star of the star-nosed mole—they take advantage of the inherent plasticity of the infant brain to make sense of the sensory information provided. One neuro-scientist sums up these relationships in one simple sentence: "The developing nervous system is also an evolving nervous system."[25]

Every undergraduate, graduate, and medical student learning about the brain must memorize the various lobes of the cerebral cortex and their particular roles in vision, hearing, touch, and smell. I certainly did. But such lessons in memorization are rarely accompanied by the caveat that our current discussion demands: that each lobe of the brain reliably can be ascribed a particular function only because the eyes, ears, fingers, and nose reliably send their neural connections to particular locales within the brain.

Thus, *our brains are not preprogrammed to expect the presence of any appendage, whether arms, legs, eyes, or stars*. Rather, it is the reliability of sensory targeting (and other developmental factors) that produces the *illusion* of a preprogrammed brain. So, if we were to examine the brain of an armless wonder, we would find a cerebral cortex that lacks map-like representations of the missing arms. In their place, we would discover enlarged representations of legs, feet, and toes.[26]

But long before an arm or a leg establishes its sensory and motor links with the brain, it exists on the embryo's body wall as a tiny protrusion, a mere assortment of cells, a limb bud. Such buds are the wellspring from which the limb develops—arm, hand, and fingers; leg, foot, and toes. The cellular processes within these buds determine whether a fully formed limb will emerge and, if so, the exact form it will take, whether that form will possess (for example) the correct number of fingers and toes, and whether it will have the wherewithal to regenerate should the original be severed. As we will see, when parasites attack a frog to produce a freak with extra limbs and digits, they target the limb bud. The limb bud has even been modi-fied to produce evolutionary novelties that we do not normally associate with limbs, such as turtle shells and beetle horns.

Limb buds have great potential.

A BUD'S LIFE

The organization of the body axis is one of the defining characteristics of any animal. For the vertebrates, which include amphibians, reptiles, birds, fish, and mammals, this axis revolves around a backbone within which runs the spinal cord and above which sits a brain. This basic plan sets the ground rules for the diversity that can be expressed.

To gain some sense of the ground rules, consider your ver-tebral column and the individual vertebrae that comprise it.

Although each vertebra may look and feel the same as the next, there are important distinctions. In particular, the varying sizes and shapes of the vertebrae reflect their structural and functional roles, which in turn reflect their location along the spine. Thus, beginning at the base of your skull, you find the relatively small *cervical* vertebrae. At your ribs, you are touching your *thoracic* vertebrae. Just below the ribs, you reach the *lumbar* vertebrae—they are larger than the others because they must support more weight. Finally, at the base of the spine, at the level of the hips, is a single bone that contains the *sacral* vertebrae. The tailbone, or coccyx, comes last.

As William Bateson made clear in his *Materials for the Study of Variation*, the spinal column can be viewed as a system that varies in two discrete dimensions: the total *number* of vertebral segments and the *identity* of a given vertebral segment. The first dimension (which Bateson called *meristic variation*) is perhaps most obvious—all other things being equal, an animal will be shorter or longer depending upon the number of vertebrae it has.

The second dimension (which Bateson called *homeotic variation*) can be seen when, for example, a vertebra in the lumbar region exhibits a form typical of a vertebra in the thoracic region. Today, *homeosis* refers to the change of a body part into one that is normally present elsewhere, as when some freakish flies develop an extra set of wings or a foot at the end of an antenna. (Decisions as to which form an appendage adopts are influenced in part by the so-called *homeobox*, or *Hox*, genes.)

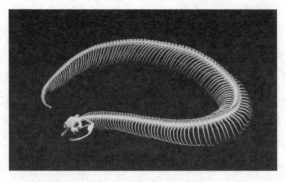

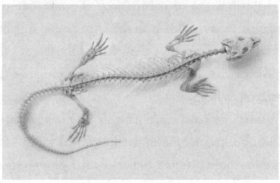

The skeleton of a snake (top). The extended rib cage, dominating the vertebral column from nearly head to tail, is clearly visible. Each pair of ribs connects to a thoracic vertebra. In the skeleton of a limbed reptile (bottom), note the limited extent of the rib cage and the placement of the limbs along the vertebral column.

Importantly, these two dimensions of variation *within* species also form the basis for evolutionary differences *among* species. Humans have thirty-three vertebrae and frogs have less than ten. Snakes are standouts, often having hundreds of vertebrae. But this serpentine abundance is only half the story, because what also distinguishes snakes from most other vertebrates is an over-commitment to one kind of vertebral element, namely, the rib-associated thoracic vertebrae. As a result, snakes are slithering ribcages. They are armless *and* legless wonders. They are as freakish as a fly with a foot attached to an antenna.

In vertebrates with appendages, the forelimbs develop from buds at the junction between the cervical and thoracic vertebrae. Similarly, the hindlimbs develop from buds at the junction between the lumbar and sacral vertebrae. At the molecular level, the expression of particular *Hox* genes during development is closely associated with the identity and organization of the vertebral elements and thus the placement of the limbs.[27] In fact, geneticists can now visualize the embryonic expression of particular genes along the vertebral column and, using this information, predict the ultimate identity of the vertebrae and their location. From these kinds of studies, scientists are piecing together the molecular and evolutionary pathways to limbs and limblessness.

Looking from the most advanced modern snakes back to their now-extinct ancestors, we can glimpse the gradations that connect them. We can see how a reptile with four limbs and the usual assortment of vertebrae was transformed, over millions of years, into the serpent that many of us fear. The dominant theme in this evolutionary story is bodily elongation—through the addition of vertebrae—and the transformation of cervical vertebrae into the multiplicity of rib-associated thoracic vertebrae. As this theme played out over time, forelimbs were the first to go, followed by the hindlimbs.

Moreover, this evolutionary path is discernible in today's snakes. Pythons are primitive snakes that, upon close inspection, have the vestiges of hindlimbs near the tip of their vertebral column—just below the ribcage. The rump end of the animal supplies enough room for hindlimb buds to take root, but they do not develop sufficiently to become functioning hindlimbs. In more recently evolved snakes, such as cobras and vipers, even these rudimentary hindlimbs are gone.[28] These snakes are completely and unequivocally limbless. What binds all snakes together is loss of forelimbs, brought about when the thorax (with ribcage) expanded toward the head, thereby occupying the space where the forelimbs of other animals are located.

How is it that a python's limb bud can take root in its proper location and even initiate the process of producing hindlimb bones, but then fail to make a fully functioning limb? In answering this question, we will gain insight not only into how limbs can be lost, but also into how limbs normally develop and how they can be modified to produce a menagerie of final forms.

The limb bud begins its life as a collection of cells at the outer edge of the embryo's body. The bud itself comprises a thin layer of outer cells—called the *apical ectodermal ridge* or AER—that encloses a clump of nonspecialized *mesenchymal* cells. During the development of a vertebrate forelimb, the bud grows outward from the body. As it does so, the mesenchymal cells differentiate into the more specialized cells that give rise to the collagen and bone of the new limb. Within each limb, bones are produced in sequence: first, the single long bone of the upper arm (the humerus), then the two bones of the forearm (the radius and ulna), and finally the bones of the wrist and digits. A parallel process occurs in vertebrate hindlimbs: first the femur, then the tibia and fibula, then the bones of the ankle and toes. The process is orderly and predictable.

Although scientists do not yet fully understand how a limb bud becomes a fully formed limb, they do have some understanding about the construction rules guiding limb development. Some of these rules are specified within the bud itself, such as in the localized collection of cells—the so-called *zone of polarizing activity* or ZPA—within the limb bud just beneath the AER. Playing a particularly important role in limb development, ZPA cells help to establish the structure and orientation of the forearm (or foreleg) and fingers (or toes). We know this from experiments in which a second ZPA is grafted into an otherwise normal limb bud, thereby stimulating the development of a duplicate limb complete with a second set of digits arranged in mirror-image fashion.[29]

For the grafted ZPA to effectively produce duplicate limb structures, it must be placed within the limb bud in direct contact with the AER. Such contact is the hallmark of *inductive* interactions, whereby abutting tissues alter each other's activity to produce an outcome that neither tissue alone can produce. It is now known that several signaling molecules mediate these inductive interactions between the ZPA and AER, including retinoic acid (a molecule closely related to vitamin A) and sonic hedgehog (the same sonic hedgehog, or Shh, that we discussed in Chapter 2). To produce duplicate limb structures without grafting an additional ZPA into a limb bud, it is sufficient to apply Shh or retinoic acid (which stimulates Shh production) to the limb bud. These and other observations have helped researchers move closer to identifying the entire cascade of molecular and physical interactions that produce a well-formed limb.

The limb bud begins in a state hovering between order and disorder. To the naked eye, it looks like a mere collection of cells contained within a skin-like outer membrane. Beneath that membrane lie the limb-making ingredients, including the ZPA hotspot. These ingredients by no means comprise a preformationist blueprint. Instead, the limb bud emerges epigenetically in developmental time through an unfolding series of events to produce limbs, which themselves do not look anything like the buds from which they came.

As someone committed to the epigenetic perspective, Pere Alberch—whom we first met in Chapter 1—advanced our knowledge of limb construction by choosing to focus on the developmental processes that produce limbs rather than focus exclusively on their final forms. He charted the journey of limb development from bud to cartilage to bone: Each bone emerges after the one before, increasingly distant from the body until the appearance of the final bone at the tip of a finger or toe. Then, equipped with maps of the forelimb and hindlimb bones for a variety of species and of the developmental trajectories for each limb, Alberch aimed to describe some basic rules of construction that might account for the observed evolutionary and developmental relationships. Finally, he tested his accounts of these relationships by experimentally manipulating the early development of limb buds in different species.

Alberch recognized that the growth of a limb is a product of local and global influences. He recognized that it is a product of physical processes that unfold through time, and of chemicals—called *morphogens*—that diffuse through the limb and alter the activity of individual cells in ways that move the growing limb along toward its final form. For many scientists, the search for these morphogens has been the search for the holy grail of limb growth. As mentioned, Shh and retinoic acid are among the molecules now known to participate in limb construction. But as was his way, Alberch charted a different course.

By painstakingly mapping the limb bones of frogs, salamanders, and other species, Alberch and his collaborators made some tantalizing discoveries. For the sake of comparison, if you hold out your right hand, palm down, in front of you, you will notice your thumb (digit 1) on the left and your pinky (digit 5) on your right. But some species have evolved more or fewer digits. In particular, possession of fewer digits, or *digital reduction*, is a pattern that has been repeated throughout evolution. Alberch was interested in *how* these digits were lost.

To understand the *how*, he first examined the *what*. He found that, in four-toed frogs, digit 1 disappeared, whereas in four-toed salamanders, it was digit 5 that disappeared.[30] Next, when he examined how toes develop in frogs and salamanders, it was evident that the ordering of toe development differs between them. Digit 1 is the last toe to appear during frog development, but the first to appear in salamanders. Thus, in a variant of the rule that the last to go in is the first to come out—a rule that you know if you have ever packed a U-Haul—Alberch hypothesized that the evolutionary loss of toes is intimately related to the developmental process that produces them.

To test his hypothesis in the laboratory, Alberch contrived to produce four-toed frogs and salamanders using related frog and salamander species that normally possess five toes.[31] He reasoned that any manipulation that produces four toes in a frog should do the same in a salamander, but that the *identity* of the missing toe would differ between species: digit 1 in frogs and digit 5 in salamanders. By injecting a chemical into the limb bud that reduces its size by reducing the number of cells comprising it, he did indeed produce four-toed creatures. As predicted, the limbs of these artificially produced frogs and salamanders resembled the four-toed variants that evolution had produced naturally. Further reduction of the number of limb bud cells produced salamander limbs with only two toes, again resembling a naturally occurring two-toed salamander.[32]

It is intriguing that such a crude manipulation—namely, using a drug that reduces the number of cells in the limb bud— does not produce a chaotic limb. On the contrary, most of the limb's parts developed where they should be, with the relatively minor exception of one or more digits. In general, there is an orderly relationship between the size of the limb bud and the number of future digits it produces.[33]

Alberch also saw clearly how this simple mechanism of digit modification—the number of cells in a limb bud—illustrates something deeper about development and evolution: If limb-bud size helps to determine the number of fingers and toes, then even a factor that only incidentally changes limb-bud size will also change the number of fingers and toes. For example, animals bred for large bodies, such as St. Bernards, Newfoundlands, and Great Pyrenees, tend to have an extra toe on each hind leg.[34]

Such incidental relations among traits place limits on the power of breeders, and even natural selection, to produce novel traits. Thus, when the developmental ties that bind traits cannot be broken—for example, if it simply is not possible to breed an animal for a large body *and* a small limb bud—then we say that development *constrains* or *biases* the evolutionary process. Put another way, developmental mechanisms load the evolutionary dice.

The flip side of this notion of developmental constraint is the notion of *interchangeability*. Just as I can take several different routes to work each morning, complex developmental processes can arrive at similar destinations despite changes to any of a number of genetic and extra-genetic factors. Alberch noted, for instance, two different ways to increase or decrease the number of cells in a limb bud: alter the number of cells that migrate into the bud or the rate at which cells proliferate within the bud. But regardless of how the number of cells is altered, a similar final limb will result.[35]

So it follows that for any distinguishable trait such as *having four toes*, we should not expect to find a "four toes gene."

To expect such a gene to exist is to misunderstand a central feature of development, namely, that every trait emerges from a multitude of local and global interactions within a network comprising numerous interchangeable parts.[36] Such is the nature of epigenesis.

As with fingers and toes, so with entire limbs. With snakes, we saw how the occupation of the vertebral column by thoracic vertebrae (and their ribs) precluded formation of forelimb buds, resulting in armlessness. But the loss of hindlegs could not be explained in this way because snakes have plenty of room for them. So where are they?

Because a limb bud is a complex brew of cells and signaling molecules, there should be many ways to disrupt its developmental path. When hindlimb buds are clearly present (and properly located), as in python embryos, it is a relatively simple matter to compare these buds to those in animals that develop hindlimbs. Such a comparison showed that python buds fail to produce the AER, the thin outer layer of tissue that encloses the cells within the bud. Without an AER and its *inductive* interactions with neighboring cells, Shh is also not produced.[37] Thus, although some of the bud's parts remain, crucial elements are missing and the cascade is blunted. Literally and figuratively, leg development has been nipped in the bud.

The evolutionary loss of limbs has happened often enough that we have the wherewithal to compare the loss mechanisms in distant groups of animals. For example, as with the ancestors of snakes, the ancestors of whales and dolphins possessed four functioning limbs that they used to move about on land. Around 40 to 50 million years ago, those ancestors began their descent into the water.[38] About 10 million years later, the transformation to aquatic life was complete, including body streamlining for movement in water, gobs of blubber to stay warm, and a fluke for propulsion. The animals' forelimbs were modified into inflexible flippers for stability and steering, whereas their unneeded hindlimbs were lost. Today, we look at these marine mammals and marvel at their adaptive elegance. But on those occasions when a whale lies helplessly on a sandy beach as its bipedal cousins struggle to heave it back into the water, it seems oddly handicapped.

The embryos of whales and dolphins, like those of snakes, possess hindlimb buds. But the buds of whales and dolphins, unlike those of snakes, transiently express an AER, thus implicating a different path to limb loss in these sea creatures. Recent findings in dolphins indicate that these buds lack the ZPA and its associated signaling molecules, resulting in the inability to maintain the AER and produce Shh.[39] If this is starting to sound like a random collection of acronyms, it is enough to remember that the limb bud, like a juggler with several balls in the air, must contend with numerous elements intertwined in space and time. Just as there are many ways to throw off a juggler—snatch a ball, cover his eyes, tap him on the shoulder—there are many ways to disrupt a limb bud.

Still, common developmental and evolutionary themes may underlie limb loss in reptiles and marine mammals. One or more bony elements comprising a rudimentary leg can be found beneath the skin of modern whales and dolphins as well as in pythons and other primitive snakes. Taking a more expansive view of the fossil record in these groups of animals, we see that the legs were lost gradually over millions of year—digits reduced, limbs shortened, fewer bones produced. A single freak mutation does not explain this history. Rather, the evolutionary path to limblessness was long, beginning perhaps with small alterations in limb bud signaling and ending with more dramatic changes in the patterning of the vertebral column.[40]

RESTORATION HARDWARE

For someone who has suddenly lost a limb, a prosthetic device can prove essential to the resumption of an active lifestyle. But would it not be nice to just grow a new one? In theory, this should be possible. That this is not possible reflects an unfortunate reality: that in addition to the limb, something else is missing. That something is a *capacity*—the capacity to regenerate.

What a wonderful trick, to lose a limb and grow a new one. We humans have some experience with regeneration. A plucked hair grows back. Lost blood is replenished. Skin cells are regularly replaced. But our limbs and those of other mammals, as well as those of birds and reptiles, do not regenerate. On the other hand, amphibians—especially newts and other salamanders—are champions of limb regeneration. Just a few weeks after amputation of a newt's limb, the full limb reforms—complete with toes, muscles, and blood vessels. In fact, salamanders can regenerate more than limbs, including the tail, retina, heart, and even spinal cord, which is why researchers continue to look to salamanders for clues to the mystery of regeneration. Resolving that mystery could dramatically improve the lives of human amputees.

The regeneration process begins soon after the removal of most—but not all—of the animal's limb. Within the first hour, healing begins with the movement of cells over the wound and the expression of particular genes. Over the next several days, the limb is prepared for regeneration as cells within the stump lose some of their adult-like characteristics, a process of *un*-development that effectively returns the cells to their embryonic state. These cells collect at the tip of the stump. From this point on, limb regeneration looks very similar to normal limb development.[41]

It is no mean feat for a salamander to restore a specialized cell to its unspecialized, embryonic state. In effect, these creatures must transform adult cells into *stem cells*, thus highlighting important areas of overlap between limb regeneration and the search for effective treatments for degenerative diseases. Indeed, these amphibians "remind us that regeneration is an ancient and fundamental biological process, and [they] challenge our creative and scientific abilities to discover how to unlock the regenerative potential within us."[42] To unlock this potential, we must trick the tissue of mammals into behaving like that of salamanders. In fact, researchers recently demonstrated that tissue from a limb-regenerating newt contains proteins that can induce mouse cells to behave in a newt-like fashion.[43]

The fact that limb regeneration, once initiated, resembles the process of normal limb development could mean that instigating stump preparation is the primary obstacle separating human amputees from the prospect of limb regeneration. But there may be other obstacles because normal limb development in salamanders exhibits a special quality that distinguishes it from normal limb development in nonregenerating animals. Specifically, amphibian limb buds are remarkable for their capacity to grow wherever they are planted. When an amphibian limb bud is experimentally attached to a novel location on the embryo—such as the head—it develops nonetheless into a perfectly respectable limb.[44] Our limbs do not do that.

What might account for this critical difference between amphibian limb buds and those of other vertebrates? The limbs of mammals, birds, and reptiles are constructed during a phase of development characterized by numerous interactions among diverse parts of the embryo. Many of the interacting parts that lie outside the limb bud are only temporarily available—once gone, they are gone forever. Thus, when a human limb bud, for example, engages in inductive interactions with these transient structures, those interactions cannot be reproduced later in life. This is a system that can make one, and only one, limb.

In contrast, the limbs of amphibians—though produced using materials that are fundamentally identical to those used in nonamphibians—do not need to find those materials outside of the limb bud.[45] For this reason, the amphibian limb bud can be described as an autonomous, self-contained unit. Interestingly, amphibian limb development also appears unique in that it is delayed to a later embryonic stage when the transient materials required by nonamphibian limb buds are no longer available. Thus, a combination of limb-bud autonomy and delayed development appears to hold the secret to this remarkable amphibian capacity—a capacity that we humans are striving to emulate.

———

THE REGENERATIVE CAPABILITIES of salamanders have not protected them against severe global threats to their existence. Not only are natural populations of salamanders—as well as frogs and toads—declining but the incidence of malformed animals is also increasing precipitously.[46] Indeed, a "grotesque spectrum of abnormalities," including extra or missing limbs and digits,[47] has been documented.

Given the sheer numbers involved—in some amphibian populations, the vast majority have exhibited these freakish limb malformations—we can safely dismiss genetic mutation as a major factor. Instead, researchers have looked to the environment. After all, amphibians possess several features that make them particularly vulnerable to environmental insult. These features include porous skin and shell-less eggs that freely allow the passage of environmental agents into the adult and embryonic circulatory systems. Like canaries in coal mines, amphibians' vulnerability makes them sensitive indicators of the declining state of our environment.

What exactly are these agents and how might they contribute to this epidemic among amphibians? There is no single culprit. Rather, research suggests that parasitic infestation is the primary cause of the limb abnormalities, and environmental contaminants—including pesticides and other pollutants—exacerbate the effects of the parasites by promoting their growth and survival or compromising the resistance of the amphibians to infestation.[48]

Establishing a clear relationship between parasitic infestation and amphibian malformation required careful detective work. Researchers needed to demonstrate in the laboratory that parasites can induce malformations similar to those seen in nature. They also needed to prove that, outside the laboratory, the amphibians in question actually come into contact with damage-producing parasites. It now appears that one such parasite—the trematode or parasitic flatworm—causes malformed limbs in salamanders, frogs, and toads.[49]

Amphibians are but one stop in the trematode life cycle. Let us begin with aquatic snails, within which the parasites reproduce asexually and prolifically. When the parasitic larvae are ready, they swim away from their snail hosts. When they meet an amphibian, they burrow into its limb bud and alter the limb that it produces. Hobbled by its distorted limbs, the amphibian host is ripe for the picking by a water bird, which ingests it along with its parasitic load. Within the bird, the trematodes reproduce again, this time sexually, creating a new crop of eggs that the bird excretes back into the water. The newly hatched parasites swim through the water until they find and infect a new batch of aquatic snails, thus completing the cycle.

Before attacking an amphibian's limb bud, the parasitic larvae form cysts on the skin surface. In tadpoles, the larvae

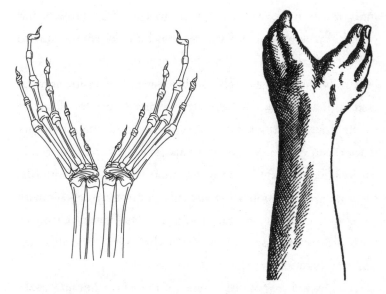

Mirror-image limb duplication in a frog (left) and human (right). The frog's condition was induced by trematode infestation; the human's condition was most likely due to a genetic mutation.

burrow beneath the skin, penetrate the hindlimb buds, and alter the spatial organization of the cells within (the forelimb buds are protected from this parasitic assault because they sit deeper inside the tadpole's body). This physical disruption alters the subsequent formation of the hindlimbs, producing malformations that can resemble those produced when experimenters surgically manipulate a limb bud.[50]

Trematode-induced limb malformations are detectable early in tadpole development and are expressed in a variety of ways, including extra limbs and mirror-image duplications. Typically, in humans, such mirror-image duplications are an inherited condition linked to a mutation that alters normal limb development. Similar conditions have been produced in genetically altered nonhuman animals. Thus, limb malformations can be caused by physical disruption or genetic manipulation, illustrating once again how epigenetic systems can arrive at similar destinations via multiple, interchangeable paths.

In frogs, trematodes can only induce limb malformations when the infestation occurs in the tadpole stage. This is not true for salamanders. On the contrary, limb malformations in salamanders can be produced at any age, thus suggesting that their capacity for limb regeneration actually puts them at risk for these malformations. In other words, the undeniable benefits of limb regeneration may have some associated costs.

To reveal these potential costs, experimenters amputated the limbs of larval long-toed salamanders and subsequently exposed them to varying levels of trematode infestation.[51] As predicted, regenerating limbs responded to parasite infestation by exhibiting a variety of malformations, most typically comprising stunted limbs but also augmented limbs and digits. Without infestation, regenerated limbs were always well formed.

However, these salamanders had one additional surprise in store for the researchers. Although the combined effect

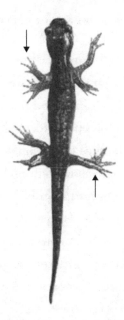

Long-toed salamander after parasite infestation combined
with experimental injury to two limbs (indicated by arrows).
Note the malformations even in the uninjured limbs.

of amputation and infection typically resulted in regenerated
limbs that were malformed, even limbs that were not amputated were malformed in the most heavily infected salamanders.
The salamander shown above, for example, displays extra limb
parts in the limbs that were amputated (and then regenerated)
as well as in those that were not. It seems, then, that parasitic
infestation of the limbs is itself injurious to the limb, thereby
inappropriately triggering the same regeneration machinery in
healthy limbs that is normally recruited to repair lost limbs.
Once regeneration machinery is activated, the parasites can
derail that machinery to produce a range of malformations.

SHELL SHOCK

It seems that anyone can appreciate just how weird turtles are. More than fifteen years ago, one of our dogs, a tiny Bichon Frise named KK, was running around the front yard when he stopped to investigate a snapping turtle. He was clearly intrigued but also very cautious—his head leaning far forward, his back legs stiff and quaking with uncertainty. Suddenly, I saw KK jerk backward, shaking and swinging his head, frantically trying to free himself from the turtle whose jaws were clamped firmly to the tip of his nose. Finally, the turtle fell to the ground, taking part of KK's nose with it. The nose would soon heal, but KK would never go near a turtle again.

As weird as turtles are, they have become so familiar to us that we may forget why they are weird. Of course, the shell has a lot to do with it. So here we focus on the shell, its unique place in evolutionary history, and its curious connection to limb buds.

About 200 million years ago, turtles suddenly (by evolutionary standards) emerged, sporting a unique four-legged body boxed within a bony shell—or more precisely, *shells* comprising the *carapace* above and the *plastron* beneath the turtle's torso. Such rapid innovations are a puzzle to biologists and laypeople alike and have traditionally inspired searches for "missing links." Such searches have since lost much of their allure: Biologists may still seek intermediate forms among ancestors, but today we understand that innovative anatomies—such as that of the turtle—can and often do arise suddenly.

Other animals, such as armadillos, use tough, scaly skin as external protection. But the turtle's protective shell stands alone in its shape and its bony composition. Why do turtles, among all four-legged animals, need this distinctive form of protection? Look closely at the turtle skeleton below and you see why: *There is not a rib in sight.*

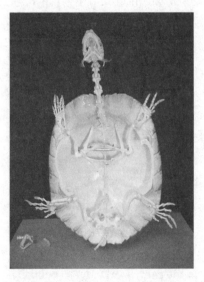

Underside view of a turtle skeleton. Note the absence
of a visible rib cage. The ribs are buried within the shell.

If we wished to weave an evolutionary story about turtles, we might conjure a tale about ancestors losing the rib cage and adding the shell as an alternative means of protection. But no one has found a turtle ancestor that lacks both ribs and shell—and no one will, because the shell represents a novel way of using ribs, not a novel response to their disappearance. Thus, the carapace embodies both cause and cure for the vulnerability of the turtle's torso.

The truth about turtles can be found in the embryo. The evolutionary innovation that is the shell begins as little more than a swelling on the embryo's flank, between the forelimb and hindlimb buds. This swelling is called the *carapacial ridge.*[52] Similar to the *apical ectodermal ridge,* the AER, that occupies the outer edge of the limb bud, the carapacial ridge also encloses mesenchymal cells that, through inductive interactions, contribute to the construction of a complex appendage.

When you and I were embryos, our ribs grew outward from the spinal column, curving downward to form the cage. Unlike our limbs, which lie *outside* our ribs, turtles' limbs lie *inside* their ribs. For this form to develop, the ribs must grow outward from the vertebral column, but never downward. In the process, they provide supportive struts for the canopy-like shell, sheltering the limbs like an open umbrella resting on a shoulder. Once again, the key to this unique relationship between ribs and limbs is the carapacial ridge. Specifically, it is the ability of signaling molecules within the carapacial ridge to divert, attract, and ensnare the ribs that sends the turtle down its distinctive developmental path. Disrupt the carapacial ridge and a nonturtle-like rib cage—one more like yours and mine—will emerge.

Beginning as only a small bud, the carapacial ridge expands outward, away from the body, and lengthwise, toward each of the other limbs. Rib cartilage inhabits the carapacial ridge and, as in other limb buds, the cartilage hardens into bone and the ridge continues to expand, ultimately forming the outer edge of the carapace. In this way, we can view the shell as a novel limb form, but one that is produced using ancient mechanisms and found parts.

The evolution of turtles did not stop with the carapacial ridge and the resultant development of the shell. The anatomy, physiology, and behavior of turtles were modified extensively to accommodate their novel form, and even the shell—although first and foremost a means of bodily protection—was recruited to aid in the storage of water and fat.

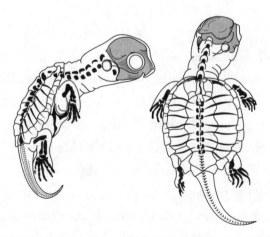

Drawings of a snapping turtle embryo highlighting the
developing ribs within the shell.

But such observations cannot detract from the claim—one that now rests on a firm empirical foundation—that the carapacial ridge was the key evolutionary innovation that set turtles apart from all of their close and distant relatives. In fact, recent research suggests that once the ridge evolved to modify the growth trajectory of the ribs, relatively straightforward molecular interactions between ribs and surrounding skin were sufficient to produce a hardened shell.[53] So turtles teach us how evolutionary novelties can arise suddenly through alterations "in the timing and positioning of common mechanisms, resulting in altered contexts for developmental events and ultimately the altered body plan."[54]

MEET THE BEETLES

In *Through The Looking Glass And What Alice Found There*, Lewis Carroll describes a fierce battle between a lion and a unicorn. During a break in the fighting, the unicorn encounters Alice, whom he regards with "the air of the deepest disgust." The unicorn exclaims, "I always thought [children] were fabulous monsters!" Alice responds in kind: "I always thought Unicorns were fabulous monsters, too. I never saw one alive before!" Then, striking a conciliatory tone, the unicorn replies: "Well, now that we have seen each other,…if you'll believe in me, I'll believe in you."[55]

If seeing is believing, then our continued fascination with the mysterious unicorn is not easily explained. Mammals

possessing single horns have never been seen in nature (the
horn of the rhinoceros is not a "true" horn as it is composed
of matted hair rather than bone). The Roman author Pliny
and others wrote of people artificially producing such creatures through various manipulations of the budding horns of
lambs or young oxen. But their writings could not erase the
considered scientific opinion that the unicorn was a creature
that neither nature nor man, beyond imagination, had ever
produced.

Franklin Dove's "fabulous unicorn."

That is where things stood until the early 1930s, when
Franklin Dove, a biologist at the University of Maine, surgically transplanted the two horn buds of a one-day-old male
calf from their usual outlying position on the skull to a single,
central location. His expectation was that "the two horns would
fuse together into one large horn solidly attached to the skull

and located between and somewhat above the eyes, as is the horn of the unicorn."[56] That is exactly what happened, and so it was that this long-sought inhabitant of myth and legend crossed into reality.

The success of Dove's experiment depended upon the ability of an isolated horn bud, like the regenerating limb of a newt, to grow independently of other tissues in the calf's body. The experiment clearly demonstrated such autonomous growth. It also highlighted the utility of producing artificial oddities through manipulations performed early in life.

Dove's accomplishment was interesting. But if we wish to learn something fundamental about horns and their relationship to development and evolution, we will do well to turn to beetles—creatures that may be less revered than unicorns but are just as exotic. In fact, when it comes to horns, beetles put unicorns to shame. This is especially true of the scarabs—one of more than 120 beetle families—which comprise tens of thousands of species and the bulk of all horned-beetle species. More significantly, scarab horns are stunningly diverse, providing biologists such as Doug Emlen with a platform upon which to explore the developmental causes and evolutionary significance of extreme variability among closely related species. As a result, the humble beetle has been propelled to the forefront of developmental evolutionary biology.

Like turtle shells, beetle horns arise through developmental mechanisms similar to those that produce limbs in other animals. The horns are rigid fixtures that emerge as the beetle

A dung beetle pupa.

undergoes late larval development. They arise from collections of cells—called *imaginal disks*—that are the insect version of limb buds. Indeed, at other places on the beetle's body, imaginal disks underlie the growth of all other major appendages, including legs, wings, and antennae. Thus, the evolution of beetle horns reflects the novel use of an established limb-production mechanism. As with the horn buds that Franklin Dove manipulated to produce his unicorn, an imaginal disk can be transplanted from one region of the beetle body to another without disrupting its ability to produce a horn, wing, or leg. In this sense, imaginal disks are self-contained organizers of insect appendages.

It is during the larval period that cells within the imaginal disks proliferate rapidly and in time produce nascent appendages. At this early stage, the developing appendages are hidden from view, "folded up and tucked away on the inside of the animal."[57] It is only as the larva transforms into a pupa, shedding its temporary larval exoskeleton, that the appendages, including the horns, unfurl to reveal the basic bodily form and features that will (unless modified during the pupal stage[58]) define the beetle's appearance for the rest of its life.

Scarab beetle species exhibit a diversity of horn size,
shape, and location.

So the larva-to-pupa transition is a beetles' coming-out party—a party whose guests look like the attractions at a beetle freak show. Consider the small assortment of scarab beetles above. Each species occupies a bizarre corner of the beetle world with its own distinct set of horn characteristics—single and double, short and long, thin and stout. The horns can emerge from the front, middle, or back of the head, as well as behind the head on the thorax.

This diversity of horn *size*, *shape*, and *location*—as well as the fact that only males, especially large males, typically exhibit horns—cries out for explanation. Clearly, given that horns emerge during the larval stage of development, *any effort to understand the evolutionary origins of that diversity must focus on the developmental mechanisms that generate horns.*

First, let us look at function. Do the particular shapes and forms of beetle horns reflect their functional use within each individual species? Perhaps beetles match their horns to particular lifestyles, just as a carpenter holding a hammer behaves differently than one holding a screwdriver. But this does not appear to be the case. Rather, male beetles simply use their horns, regardless of their location and shape, as weapons to fight other males for prime territory—whether at entrances to tunnels, on bamboo shoots, or near sap leaking from trees—anywhere female beetles can be found. A successful battle for access to females provides the opportunity to win the reproductive war.

Dung beetles provide a particularly vivid example of how this social system works. The larvae of dung beetles develop within a tunnel that their mother excavates. This tunnel is built beneath piles of dung, a housing arrangement that provides a ready source of food for the mother and her offspring. As egg-laying time approaches, any male that gains access to the mouth of the tunnel and fends off all other suitors will be in position to fertilize that female's eggs, and the single most important attribute that determines a male beetle's success in this endeavor is his size. In the world of beetles, it is good to be a large male with long horns (although we will see in the next chapter that small, hornless males compensate for their deficiencies in other ways).

Now that we know that size matters, let us look at the processes that establish a beetle's size and examine the relationship between body size and horn length in a random collection of male beetles. From this simple analysis we quickly find that each species has a threshold body size that separates, so to speak, the men from the boys. Below that threshold, males are both small and hornless. Above that threshold, as bodies get bigger horns get longer. So it appears that a *developmental switch* generates two sub-populations of male beetles: The horned and the hornless. The haves and the have-nots.

When we see such distinct groups of individuals, we may be tempted to imagine genetically distinct populations. We should resist this temptation. In ants, the elaborate caste system composed of sterile workers and queens can seem an impossible creation without the action of distinct "caste genes." But these caste members, despite their distinct bodies and behavior, are genetically indistinct. Rather than genetic inheritance, it is developmental experience—including such environmental factors as nutrition, temperature, and chemical signals produced by other ants—that shapes each ant's destiny.[59]

So it is with beetles: The nutritional environment of each larva is a critical factor in determining its final body size and horn length. For example, in dung beetles, when the eggs are laid, each egg is partnered with a clump of dung—egg and dung together forming a *brood ball*—that the mother previously placed deep within the tunnel. Her male mate may have helped to supply the tunnel with dung; in fact, paternal help can have a significant impact on the growth of larvae.[60] Bigger brood balls mean bigger dung beetles. Also, because mothers working with fathers are better providers of dung to developing larvae than mothers working alone, and because helping fathers are always big and horned, the result is that big fathers tend to produce big male offspring[61]—a nice example of how an evolutionarily important character can be inherited by the next generation without the mediation of a genetic mechanism. Such examples of nongenetic transmission of characters are now becoming commonplace and are helping to solidify the notion that the heredity upon which evolution depends is more than just about genes.[62]

Next, we must explain the relationship between big bodies and long horns. In that regard, it might seem reasonable to expect that big males have long horns for the same reason that elephants have bigger heads than we do. But there is more to this story. The link between beetle body size and horn length can be found in the secretion of several signaling molecules—including juvenile hormone (JH) and insulin—that are produced in greater amounts in larger males. These hormones interact to regulate the rate and duration of cell proliferation in the imaginal disk that produces a horn.

Thus, well-fed larvae will grow bigger bodies and their imaginal disks will be bathed in higher levels of JH and insulin,

resulting in imaginal disks that exhibit faster cell proliferation for longer periods, thus producing longer horns. This
link between nutrition, body size, and hormone levels helps to
establish the orderly relationship between body size and horn
size that researchers have documented in their beetle subjects.
This orderly relationship is not fixed in stone, however, for
small males can be tricked into producing long horns merely by
exposing them to artificially high levels of JH.

But of course, signals are only meaningful when there are
receivers to detect them. Television stations are of little use
without televisions to receive their signals. Similarly, with hormones like JH, there must be JH *receptors* to translate the presence of the hormone into meaningful biological activity. Each
receptor is a protein molecule, typically embedded within the
membrane of a cell, that responds to the presence of a specific
hormone just as a lock responds to the insertion of a specific
key. Hormones and receptors work in concert to produce
downstream biological effects.

Moreover, because hormones float freely throughout the
bloodstream, evolutionary modifications to specific structures
(like limbs and horns) must often be achieved through developmental modifications of hormone receptors, including their
location, quantity, and sensitivity. Merely changing the quantity or timing of hormone release would not be effective for
producing *specific* changes in horn size and location because any
change in hormone levels will change a myriad of other structures that are also responsive to those hormones. Thus, based

on these basic principles of hormone function, we should expect the size and location of beetle horns to track closely the number and location of horn-producing cells equipped with receptors for such hormones as JH or insulin.[63]

Accordingly, many such hormone-responsive cells in a central location on the head would, in the presence of sufficient hormone, produce a single large horn protruding from the head. Conversely, when horns are absent, as with females and small males, we would expect to find fewer such hormone-responsive cells. These expectations have yet to be confirmed experimentally. Regardless, as Emlen points out, it is clear that an understanding of developmental mechanisms steers us toward evolutionary insight: "By learning how horn expression is regulated, we can begin to think about how horn expression might be modified...[and] begin to think about what types of modifications are likely to have occurred in the past to generate the diversity of forms that exists today."[64]

―――――

BEETLE HORNS ARE not only about benefits. They also have costs, which provide these odd little creatures with one more opportunity to teach us something fundamental about development, evolution, and diversity.

The costs associated with producing beetle horns come in several forms. As with the construction of any appendage, horn construction requires time and energy. The sheer heft of some horns means that sizable quantities of energy must be allocated to horn growth during a period of development when other growing tissues also need energy. Robbing Peter to pay Paul has its consequences. So does producing horns that can be so large that they hinder a beetle's freedom of movement, exposing him to dangers that do not confront beetles with more modest endowments. But we can be reasonably certain that the benefits of horns to large males—that is, as weapons that enhance a male's access to females—are sufficient to offset the associated costs.

We have focused much of our attention in this chapter on buds and the processes that transform a simple collection of cells into limbs, shells, and horns. But as we consider the costs of horn growth in beetles, we see that there is more to limb development than limb buds. Buds live and grow in neighborhoods, and their neighbors have needs, too.

In some beetle species, horns grow toward the back of the head—that is, in the same neighborhood as the eyes. In such species, males with large horns have small eyes, and males with small horns have large eyes. These inverse relations between anatomical features could be due to horns and eyes responding in opposite but independent ways to the same developmental environment. Alternatively, such relations could arise from horns and eyes *competing* with each other for the same developmental resources within the same small space on a beetle's head.

This latter explanation appears to be correct, as a variety of experimental manipulations that target horn length—for example, administering JH to male larvae—produce beetles with longer horns and smaller eyes. What is difficult to produce is a beetle with longer horns and larger eyes, or shorter horns and smaller eyes.[65] It seems that the cost of horn growth in these beetles is literally written on their faces.

This insight applies to more than just eyes. Beetle species that grow horns at the front of the head, in the same neighborhood as the antennae, exhibit a trade-off between horn length and antenna size; and beetle species that grow horns on the thorax, in the same neighborhood as the wings, exhibit a trade-off between horn length and wing size. So the growth of a horn constrains the growth of adjacent appendages, regardless of what those appendages are.

This observation should not surprise us. Two hundred years ago, Etienne Geoffrey Saint-Hilaire noted that if "one organ takes on extraordinary growth, the influence on neighboring parts can be observed: they then no longer arrive at their usual development."[66] Similarly, nearly a century ago, Charles Stockard identified the "study of the growth influences of one embryonic organ on another [as] one of the most important problems in the analysis of structure."[67]

Having demonstrated the effect of horn growth on neighboring appendages, Emlen went further, relating beetle-horn location—one of the hallmarks of beetle diversity—to the lifestyles of individual species. He predicted that beetle species that relied heavily on vision to move about their world—such as those that are most active at night when large eyes with heightened sensitivity to light are particularly important—should be less likely to have large horns at the base of the head. He reasoned that the development of large horns there would compete with the growth of the eyes and produce a visually impaired beetle. In fact, when he tested this idea by examining scores of beetle species, he found that night-active beetles were less likely to have large horns at the base of the head, thereby freeing up the eyes to grow sufficiently large to meet their nocturnal needs.[68] In this way, the location of a beetle's horns tells us a lot about its lifestyle.

With this observation, Emlen moved beyond the issue of competitive growth to say something deep about the evolutionary forces generating diversity. But his insight required more than a simple cataloguing of traits. When other researchers produce such catalogues—without information about the ease or difficulty of developing various traits and about the competitive interactions among developing structures—they are often tempted to "explain" each trait as independent productions of trait-specific genes. Such simplistic invocations of genetic design and control rarely move our science forward.

In contrast, our examination of beetles demonstrates how much more we can learn about evolution through careful analysis of *all* the mechanisms and circumstances of development. "Only by beginning to explore aspects of the development of a trait," Emlen writes, echoing Pere Alberch before him, "can we begin to identify the *causes* of bias or correlation among traits."[69]

We have not yet achieved the *theory of form* that Alberch dreamed about, a theory where "the properties of interactions that characterize development" take precedence over any "specific genetic constitution."[70] The critical elements that any such theory must contain, however, are clearer today than they have ever been: Complex networks comprising local and global molecular signals, autonomy and interdependence, constraint and bias, costs and benefits, competition for growth, and interchangeable influences all help to explain how limbs and other appendages are made, lost, misplaced, regenerated, and transformed.

More generally, these elements have all contributed to the evolutionary production of diverse and innovative structures. With so many illuminating examples of how complex interacting networks conspire to produce bodies and brains, we are beginning to understand why some innovations are so easy to achieve and some are so hard. Building on this insight, we are marching steadily toward an accurate and durable conception of the developmental origins of evolutionary change.

FIVE

ANYTHING GOES

When it Comes to Sex, Expect Ambiguity

> *My change from girl to boy was far less*
> *dramatic than the distance anybody*
> *travels from infancy to adulthood.*
> JEFFREY EUGENIDES, MIDDLESEX[1]

ere Alberch worked hard during his all-too-brief lifetime to unify evolutionary and developmental biology. Today, this unification seems within reach, as scientists—under the banner of Evo Devo—struggle to unravel the evolution of embryonic development. Would Alberch be content with the direction of these current efforts? I asked this question of someone who would know the answer. Ann Burke was a graduate student in the 1980s who worked with Alberch at Harvard University. Today, like her former mentor, she studies the developmental processes that underlie evolutionary change.[2]

Alberch, Burke said, would have been disappointed (although perhaps not surprised) with today's continued preoccupation with genes at the expense of developmental systems. It is a disappointment she herself experiences—a frustration with the "hijacking" of her field "by a *genocentric* attitude" fueled, she believes, by an intolerance for ambiguity.[3]

Ambiguity unsettles us. Consequently, we turn to dichotomies to help us divide the world into neat, unambiguous categories: good versus evil, genes versus environment, liberal versus conservative, even genocentric versus nongenocentric. But of all dichotomies, perhaps none is more entrenched, more personal, and more vigorously defended than that of male versus female. With each new challenge to the sanctity of this dichotomy—typically coming from individuals or groups seeking to fit into a culture that seems unable or unwilling to accept them—a wave of reactionary responses ensues to "save" our social and legal institutions from ruin. As each wave passes, partisans on both sides regroup and, like medieval armies, suit up for the next battle. And so it goes.

In the battle over the sexes, ambiguity takes many forms. Perhaps most familiar to us are those ambiguities that arise from seeming mismatches between behavior and anatomy, such as men who love men and women who love women; or men who prefer to dress like women, and women, like men. But ambiguity can also apply to the genitalia, as with penises that are too small or clitorises that are too large, vaginas paired with testicles, or penises that accompany ovaries and breasts.

For many, such anatomical ambiguities can be so disturbing—the offending genitalia of these *intersexes* so unsettling—that knives and needles are often hastily recruited to correct them surgically. Thus, among the victims of these procedures are the ethical standards that are supposed to guide medical practice. Indeed, many doctors apparently believe that some of nature's errors are so grotesque that they are morally justified in using any means necessary to fix them. In a world guided by such so-called *monster ethics*,[4] subjecting a patient to unproven procedures is a small price to pay for the moral clarity provided by unambiguous genitalia.

As we have now seen many times in this book, embryonic development is a risky business. And in the pantheon of things that can go wrong, an ambiguous sex assignment can be as disturbing as any conceivable developmental anomaly. So in response to the distress and uncertainty that parents under-standably feel when their otherwise beautiful baby straddles the gender line, doctors have made available to them everything that the fields of urology, cosmetic surgery, endocrinology, and psychiatry have to offer. As the ambiguous baby sleeps peace-fully in his crib, a team of doctors quickly gathers the tools needed to finish the job that nature started but failed to finish. Like Arthur Goldman introducing The Bionic Man, you can almost hear these doctors saying: "We can rebuild him. We have the technology."

Intersex children do not immediately experience the distress that their parents feel. It takes time to learn that you do not fit in. None of us knows instinctively how we are supposed to look or act. We learn about ourselves through self-exploration and through comparison and interaction with others. If and when we feel distress about some aspect of our selves, it is because others make us feel that way. They point out differences. They ridicule and reject.

This intermingling of sex, medicine, and shame was effectively brought to life in Jeffrey Eugenides' coming-of-age novel, *Middlesex*. The novel's narrator is Calliope Stephanides (Callie, for short), a child who has inherited a rare mutation that prevents the chemical conversion of testosterone to a related form that is particularly important for the growth of the embryonic penis. This mutation results in a penis that is so ambiguous that Callie's parents raise her as a daughter.

Callie's form of intersexuality is known as *5-alpha-reductase deficiency syndrome*, a condition that is characterized by a range of genital forms—from the feminine to the masculine. With the arrival of puberty, surging levels of testosterone usually resolve any ambiguity in favor of a penis. But in Calliope's case, her diminutive penis was also *hypospadic*, that is, the urethra opened on the underside of the penis rather than at the tip.

In one of the novel's more harrowing scenes, the fourteen-year-old Callie has traveled with her parents from Detroit to New York to consult with an expert, Dr. Luce. When she overhears a group of doctors discussing her case, she decides to educate herself at the New York Public Library. There, with a dictionary in front of her, she looks up the definition of *hypospadias*, then follows the trail of synonyms through *eunuch*, then *hermaphrodite*, and then—to her horror—*monster*. The intervening words do not cushion the blow:

> The synonym was official, authoritative; it was the verdict that the culture gave on a person like her. *Monster*. That was what she was. That was what Dr. Luce and his colleagues had been

saying. It explained so much, really. It explained her mother crying in the next room…. It explained why her parents had brought her to New York, so that the doctors could work in secret. It explained the photographs, too. What did people do when they came upon Bigfoot or the Loch Ness Monster? They tried to get a picture. For a second Callie saw herself that way. As a lumbering, shaggy creature pausing at the edge of the woods.[5]

What she is told next, however, confuses her even more. According to Dr. Luce, Callie was "a girl whose clitoris was merely larger than those of other girls."[6] All she needs, he tells her, is some reduction surgery—to "finish her genitalia"—and a regimen of hormone therapy. But this information does not jibe with hypospadias, that is, a malformation of the *penis*. So which was it, enlarged clitoris or malformed penis?

Later, in Dr. Luce's office, she inadvertently eyes her medical report on his desk. Now she discovers the unvarnished truth. According to the report, she is a male—genetically, gonadally male—with a feminine demeanor and identity. To prevent any further confusion and humiliation, and to make marriage (if not reproduction) possible, the report advises that she undergo a "feminizing surgery" that, among other things, includes the removal of her undescended testicles and the creation of an appropriately sized clitoris. Then comes the final blow, the proof of the monster ethics at play: "Though it is possible that the surgery may result in partial or total loss of erotosexual sensation," the report concludes, "sexual pleasure is only one factor in a happy life."[7]

Rather than subject herself to these treatments and procedures, she escapes. She does so with the knowledge that Dr. Luce's dishonesty with her was matched to some degree by her own dishonesty with him; that she had not been entirely forthright with him about her discomfort living life as a girl. So she escapes, but not before declaring to her parents that she is a boy.

The transformation from Callie to Cal has begun. But this transformation cannot, as one can imagine, be easy. As Cal later admits, "I never felt out of place being a girl. I still don't feel entirely at home among men."[8] Cal must accept that he doesn't quite fit in—either into the gender molds that our culture makes available to shape him or into the scientific theories crafted to explain him. In the end, he's just different. He is who he has become. In Cal's words: "Biology gives you a brain. Life turns it into a mind."[9]

In his novel, Eugenides has accurately depicted the central concerns and conflicts that face intersexes, their parents, and their doctors. For their part, parents must grapple with the reality that their intersexed child lives in a society in which the appearance of one's genitals is a springboard to one's identity and behavior. In 1969, two physicians voiced their empathy for the parents in this way:

One can only attempt to imagine the anguish of the parents. That a newborn should have a deformity…[affecting] so fundamental an issue as the very sex of the child…is a tragic event which immediately conjures up visions of a hopeless psychological misfit doomed to live always as a sexual freak in loneliness and frustration.[10]

It is this sense of doom that motivates doctors to pull out all stops to bring these infantile sex organs into alignment with our expectations. Indeed, on its face, such efforts are no less reasonable than similar efforts to repair other congenital deformities, from defective hearts to cleft lips. Once we know how hearts and lips are supposed to function and how they are supposed to look, the steps to fixing them are relatively straightforward.

But genitalia are very different because when a surgeon repairs the genitalia, she must first decide what constitutes a fix. That decision, in turn, is determined by whether the surgeon believes she is dealing with a large clitoris or a small penis— and whether the surgeon believes a newborn baby can adjust with equal alacrity to any surgical decision. And those beliefs are intimately tied to our understanding of the definitions and developmental origins of maleness and femaleness.

What determines maleness and femaleness? Our lay definitions are sufficient most of the time—that, at the most basic level, males possess XY chromosomes and females possess XX chromosomes—but in fact, even that simple association carries a host of ambiguities. The International Olympic Committee (IOC) discovered this for itself after several decades of mandatory screening of all female athletes to ensure the "purity" of gender-segregated competition. After genetic testing replaced the humiliation of the strip search, the IOC discovered what experts in the field had long known: that having a male-typical XY genetic constitution is not incompatible with femininity. As a consequence, in 2000 the IOC officially abandoned its quest for the pure female athlete. As the chair of the IOC's medical commission recently remarked, "We found there is no scientifically sound lab-based technique that can differentiate between man and woman."[11]

It is the nature of lawyers to draw lines and make unambiguous distinctions. But that is not the nature of nature. Thus, instead of looking at men and women as mutually exclusive, nonoverlapping dichotomies, scientists have adopted a more expansive view. Indeed, we inch closer to reality by thinking of sex as a "syndrome," a collection of "symptoms" that, as a collective, allow for a "diagnosis" of *male* or *female*. Genes, gonads, genitals, and behavior must cohere—which they typically do—to produce an unambiguous individual. But ambiguities arise because this coherence is not preordained. They arise because our final gendered form is constructed in real time. They arise because that construction can be diverted and distorted in a multitude of ways—ways that reflect the presence of real-world mechanisms at work.

Of course, it is conceivable that the two sexes could have evolved as true dichotomies separated by unbreachable walls. It is conceivable, for example, that a single genetic embryonic decision could have taken care of everything at once—gonads, genitals, and behavior—such that all are always perfectly harmonized and synchronized, thereby rendering ambiguities impossible. But that is not how nature works. That is not how we evolved. That is not how we develop.

When our sexual anatomy, physiology, and behavior hang together to form a single narrative, our sexual world seems simple. But Callie Stephanides and her nonfiction human counterparts do not present us with simple stories. So we can argue about Callie's "true" sex, or we can—like a doctor who must break a bone to set it properly—shatter our preconceptions and take a fresh look at this all-too-familiar problem. As we will see in this chapter, only an ecumenical approach to sex can accommodate the human condition as well as the full range of sexual diversity in the animal kingdom. If God had anything to do with this diversity, then He clearly has a taste for deviance.

A MEMBER OF INFLUENCE

We have all seen it. A seemingly innocuous discussion heats up and begins to boil, the two combatants circling each other like wild animals until one of them, his back against the rhetorical wall (my wife assures me that this is usually a man), explodes with rage and physical threat. This threat of physical force effectively ends the argument in a seeming victory for the aggressor. But such victories are not legitimate because arguments cannot be validly won using tactics that bear no relation to logic or truth.

Logicians have a name for every imaginable argument,[12] even one where an individual steps outside the range of calm, rational behavior and resorts to physical force. It is called, in the language preferred for such things, *argumentum ad baculum*—meaning literally *an appeal to a big stick.* That stick can come in many literal and figurative forms. For example, when one is encouraged to accept the Almighty as one's personal savior or face an eternity in hell, that is an example of *argumentum ad baculum*. An eternity in hell is a very big stick.

Baculum is also the technical name for the penis bone, or *os penis.* Most male mammals, including dogs, bears, and walruses, have this bone and so never suffer the indignities of erectile dysfunction because their erections are, in a sense, permanent (females of these species typically have a smaller version of the bone, the *os clitoridis*). Humans, on the other hand, are not intimately familiar with these bones—that is, beyond the colloquial description of erections as *boners*—because we are among the minority of mammals that do not possess them. Human males must rely exclusively on a relatively fragile system of fluid hydraulics, in which the penis is swollen with blood to achieve an erection. The bane of many males has become a boon to the pharmaceutical industry.

A recent article in the *American Journal of Medical Genetics* suggested that erectile dysfunction results from an underlying disorder that afflicts every human male. The authors of this article call the disorder *congenital human baculum deficiency*.[13] Although these authors are aiming for humor, they are also making a serious point of (literally) biblical proportions. Their point rests on the reasonable assumption that males of biblical times, like their modern descendents, might have noticed how their penises sized up to those of others, including other animals. (This would have been especially true for anyone who spied the thirty-inch baculum of a walrus.) Thus, people might have noticed the prevalence of penis bones in animals including dogs, cats, and monkeys, and might have been motivated to create an explanatory myth to account for the human male's peculiar penis.

Such a myth, according to the article's authors—one a developmental biologist and the other a biblical scholar—can be found in the second chapter of the book of Genesis. You may recall that God is said to have created Eve from Adam's *rib*, a word that, in Hebrew, could refer to an actual rib or to any support structure (such as a beam). Because there is no word in Biblical Hebrew for *penis*, if the authors of the *Bible* wished to refer to that body part, they would have been forced to rely on indirect language. Moreover, given that the rib, in contrast with the penis, can claim no obvious literal or symbolic association with the act of creation, and that males and females share the same total number of ribs (as the *Bible*'s authors were certainly aware), the suggestion that the Creator might have borrowed Adam's baculum to engender Eve makes consummate sense.

With Adam's "rib" reinterpreted as a baculum, the book of Genesis continues its fascination with the penis in Chapter 17. In that chapter, God informs Abraham that all male offspring, at eight days of age, are to be circumcised or be cut off from God's covenant with the Jewish people. Thus began that oddest of traditions where family and friends gather together, recite prayers, and conspire in an ancient act of genital defacement—an act performed by a man who has made slicing the infant foreskin his hobby (or worse, his profession).[14]

With so many circumcisions performed each year by physicians and non-physicians alike, it is inevitable that some will be botched. Perhaps the most famous botched circumcision, occurring in the mid-1960s, inspired one of the most significant debates of our time regarding the pliability of human sexual identity. A careless physician, opting for a cautery rather than a scalpel, accidentally burned off the penis of a Canadian boy undergoing a circumcision at 8 months of age. David Reimer, the unfortunate infant, was then taken to John Money, a preeminent sex expert at Johns Hopkins University, for an assessment of what could be done to help the child adjust to the damage that he had sustained.

Money had become world-famous for his theory concerning sexual development, a theory that can be encapsulated in this way: Although a human newborn typically exhibits sexually unambiguous genitalia, his/her sexual identity is neutral at birth and can develop in either the male or female direction as long as the genitals "look right" and the infant's parents behave toward the child with consistency and conviction.[15]

Accordingly, Money prevailed upon David's parents to raise him as Brenda. This "reassignment"—which included removal of the child's testes, the surgical creation of a vagina, and hormone treatments—was hailed as a success until two researchers, dubious of Money's claims, followed up on the case many years later and demonstrated convincingly that Reimer had never accepted his reassignment as a female.[16] In fact, at fourteen years of age, Reimer began the process of reassigning himself back to a male. Later, as an adult, he married a woman, but after a lifetime of trauma, depression, and drug use, he took his own life in 2004 at the age of thirty-eight.[17]

Because David Reimer's circumcision and the ensuing reassignment happened many months after his birth, it is possible that the delay between birth and reassignment doomed his chances of accepting himself as a girl. If so, then Reimer's tragic experience with sex reassignment did not provide a strong refutation of Money's theory, which was formulated in terms of neutral sexual identity *at birth*. In any case, when I was a graduate student in the 1980s, Money was a celebrated scientist whose views were widely accepted. But by the time of his death in 2006, he had witnessed the unraveling of his reputation and his life's work.

The most formidable challenge to Money's ideas about sexual identity eventually came from within his own institution. In 1975, Paul McHugh—notable throughout his career for a firm commitment to psychiatric practices based on clear thinking and sound empiricism—became psychiatrist-in-chief at Johns Hopkins Hospital and assumed responsibility for all practices and procedures in his psychiatry department. As he became increasingly familiar with one particular patient group that Money had studied at length—adult males with gender identity disorder (they are usually male) who, reporting that that they feel "like a woman trapped inside a man's body," seek surgical and hormonal treatments to alter their sex—he began to doubt the prevailing view, nurtured by Money, that these were simply women bundled with the wrong anatomy. McHugh observed, for example, that although these transsexuals ached to be perceived as women, they seemed only to be acting the part. As one of his female colleagues reportedly remarked after meeting with one of these patients, "Gals know gals, and that's a guy."[18]

McHugh became increasingly uncomfortable with what he saw as a fashionable treatment that "did not derive from critical reasoning or thoughtful assessments."[19] He wondered whether the transsexual's

> sense that he is a woman trapped in a man's body differs from the feelings of a patient with anorexia nervosa that she is obese despite her emaciated…state. We don't do liposuction on anorexics. So why amputate the genitals of these patients? Surely, the fault is in the mind, not the member.[20]

McHugh believes that those individuals identified as transsexuals fall into two separate and very different groups. One group comprises homosexual men who feel shame over their orientation and see sex reassignment "as a way to resolve their conflicts over homosexuality by allowing them to behave sexually as females with men."[21] The second group comprises heterosexual men who are sexually aroused by the act of cross-dressing; for these men, surgical sex transformation carries this obsession to its logical extreme. (Because these newly transformed men retain their sexual interest in women, they also declare themselves lesbians.) Accordingly, these conflicts and obsessions reflect psychological difficulties that cannot be excised with a knife.

Regardless of the cause of transsexuality, an overriding concern for any physician is whether the prescribed treatment works, and it is true that when transsexuals were assessed years after their genital surgery, most of them did not regret their decision. Nonetheless, according to McHugh,

> in every other respect, they were little changed in their psychological condition. They had much the same problems with relationships, work, and emotions as before. The hope that they would emerge now from their emotional difficulties to flourish psychologically had not been fulfilled.[22]

McHugh concluded that any hospital that performed such surgery "was fundamentally cooperating with a mental illness."[23] As a consequence—and despite the fact that Hopkins had become the spiritual home of sex-reassignment surgery—McHugh decided that these operations for transexuals would no longer be performed at his hospital. Other hospitals soon followed his lead, although many throughout the world remain willing to conduct these surgeries.

INTER ALIA

Transsexuality and intersexuality are sometimes viewed as variations on a single theme, especially in the popular press.[24] But there are sound reasons for keeping them separate. For example, transsexuals differ from intersexuals in having unambiguous genitalia. Such a difference alone might reasonably justify McHugh's skeptical attitude toward "corrective" surgery for transsexuals. But even for the one to two percent of all live births that are classified as intersex,[25] there is today a growing movement away from hasty surgical interventions—especially for newborn infants who are incapable of consenting to a decision that will affect the rest of their lives.[26]

There are three broadly recognized classes of intersexuality in humans:

- "True hermaphrodites": These people usually possess the XX chromosomal pattern and, for unknown reasons, have gonads comprising both ovarian and testicular cells. Their genitalia can range along a continuum from typical male to typical female. Relatively little is known of the causes and consequences of this condition because it is rare and difficult to detect (to confirm this condition, gonadal tissue must be examined under a microscope).

- "Male pseudohermaphrodites": These people possess the XY chromosomal pattern and testes. Callie Stephanides, the protagonist of *Middlesex*, fell into this category. As mentioned earlier, her condition is called *5-alpha-reductase deficiency syndrome*. In the absence of the enzyme, 5-alpha-reductase, testosterone cannot be converted to dihydrotestosterone and, as a consequence, the genitalia are incompletely masculinized at birth. This condition is relatively common at certain places throughout the world, including the Dominican Republic and Papua New Guinea.

A second major form of male pseudohermaphriditism is the condition known as *androgen insensitivity syndrome* (AIS). In this condition, testes are formed and testosterone is produced, but an inability to produce testosterone *receptors* renders the testosterone useless. As a consequence, these genetic and gonadal males are raised as females, identify as females, look like females, and behave as females.

Other variations within this category involve, in otherwise typical males, diminished development of the penis, including hypospadias (where the urethra opens somewhere along the underside of the penis rather that at the tip) and micropenis.

- "Female pseudohermaphrodites": People in this group possess the XX chromosomal pattern and ovaries. However, their external genitalia are masculinized due to prenatal exposure to high androgen levels most commonly produced by the fetal adrenal gland, a condition known as *congenital adrenal hyperplasia* (CAH). Other sources of androgens include tumors in the mother and environmental agents that mimic their effects. This condition results in an enlarged clitoris and occasionally labia that are fused to resemble a scrotum. The internal genitalia, however, are generally unaffected.

As long as John Money's perspective—that a newborn's sexual identity remains open to manipulation for some time after birth—was considered correct, physicians deciding on a course of treatment for an intersexed infant were not bound by the infant's genetic, gonadal, or genital sex. Accordingly, a protocol was established whereby, within forty-eight hours after birth, doctors were to settle on a sex assignment and, if necessary, a course of surgical action.

According to accepted protocol, XX infants were to be raised as girls, this decision resting on the desire to retain her capacity to bear children. For those with CAH and other intersex conditions characterized by masculinized external genitalia, a surgeon would feminize the clitoris, enlarge the vagina, unfuse fused labia, and perform any other procedures necessary to ensure a pleasing visual experience. Unfortunately, as with the genital cutting forced upon young African girls, this surgery destroys the nerve endings in the clitoris. But many parents and surgeons accepted this cost rather than compel a young girl to endure the "shame" of a large clitoris. Moreover, according to Money's perspective, a little girl with a large clitoris would find it difficult to settle comfortably into a female identity. Therefore, surgical disambiguation of the genitalia became the standard treatment.

For XY boys, the accepted protocol was more extreme. First and foremost, it was agreed that for a newborn boy to remain a boy, he must possess a phallus with a prayer of becoming a proper penis. For that prayer to be answered, doctors decided that the newborn's penis must reach a length of at least 2.5 centimeters. If it did not, then the penis was to be summarily removed, the testes extracted, a vagina created (from a piece of intestine no less), and labia sculpted. John was to become Joan.

This decision follows a simple course of deduction: a male infant with a puny penis cannot become a true male; it is much easier to create a vagina than it is to extend a penis; and genital reconstruction at birth coupled with parental counseling is sufficient to produce a convinced and convincing—albeit sterile and sexually desensitized—female. As it turns out, each of the pieces of this deduction is incorrect or at best dubious.

The notion that vaginas are easy to create derives from a misunderstanding of what vaginas do. True, penises are complicated devices with elegant control over erection, urination, and ejaculation. Extending the length of such a device is no simple task. But the notion that a vagina molded from a piece of intestine is adequate—and is to be preferred to a small but sensitive penis—indicates a lack of respect for the vagina as a self-lubricating, shape-shifting, and—perhaps most importantly—sensitive organ. Thus, when surgeons have considered the possibility of constructing vaginas for their patients, their definition of success has entailed little more than the creation of "a receptive hole."[27]

Similarly, when surgeons aim to fix a malformed penis, they often do more harm than good. For example, the process of fixing a hypospadic penis—that is, reconstructing the urethra so that it opens at the tip of the penis—can include years of painful surgeries that ultimately leaves an otherwise functional phallus in a scarred and dysfunctional state. As with the legless child, discussed in the previous chapter, who was "helped" by placing her in a steel and leather contraption, surgical interventions with hypospadic penises can have a crippling effect.[28]

Perhaps, then, it is advisable to refrain from surgery to correct a hypospadic penis, at least until the child is old enough to consent to that course of action. But what should be done about a boy with a micropenis? We have seen throughout this book how the growing embryo and infant can accommodate insults and interventions more easily than a fully grown adult. So, would it not be advisable to commence sex reassignment, including surgical removal of the micropenis, as early as possible so as to give the developing child the best chance to live a "normal" life? The answer would be *yes* but for two problems.

First, contrary to conventional wisdom, an adult male with a micropenis is not doomed to sexual oblivion. Indeed, in one study of twenty heterosexual adult males with micropenises that were not surgically altered, fifteen had experienced successful sexual intercourse with women:

[T]he most surprising feature of these patients was the firmness with which they were established in the male role and the success that they had in sexual relationships....The partnerships were stable and long-lasting, a situation that some patients attributed to the extra attention that had to be paid to intercourse because of the short penis. Although vaginal penetration was usual, there was an *experimental attitude* to positions and methods. One patient was the father of a child.[29]

Like Johnny Eck recruiting his arms to compensate for a legless life, these individuals with small penises adapted and adjusted to their predicament. This should not surprise us. After all, are not the early sexual gropings and fumblings of boys and girls with typically sized and structured genitalia indicative of an identical "experimental attitude"? All of us must learn how our particular bodies work. We are not hardwired for sex and there is no innate instruction manual. Getting sex right, for our partners and us, entails many trials and a lot of errors. As with any learning process, feedback is essential. Slicing away at sexual organs to produce a cosmetic improvement severs communication between genitals and brain, thereby thwarting the very learning process that makes the development of sexual behavior possible.

Second, accumulating evidence directly contradicts the entrenched view, inherited from John Money, that the sexual identity of human infants at birth is neutral and therefore malleable. David Reimer's tragic case cast doubt on but could not unquestionably dispute this view because the sex reassignment following his botched circumcision occurred so many months after he was born. Reassignments of other infants at or shortly after birth have, however, more clearly argued against the validity of Money's theory. Perhaps the strongest such evidence comes from a recent study of 16 XY males with a congenital defect known as *cloacal exstrophy*, a condition that affects 1 in 400,000 male and female infants.[30] Although technically not considered an intersex condition, cloacal exstrophy entails abnormal development of the entire pelvic region that, for males, includes the near or complete absence of the penis.

As with micropenis, the standard practice for males with cloacal exstrophy was to transition affected males to the female sex (of course, females with this condition are never reassigned). Accordingly, of the sixteen male subjects in this study, fourteen were reassigned as females within weeks of birth by removing the testes and surgically producing female-like genitalia. Two of the remaining sixteen patients were raised as males because their parents would not consent to reassignment. When all these children were later examined to assess their degree of comfort with their sexual identity, they were between eight and twenty-one years of age.

Despite sex reassignment and parental cooperation in raising their male offspring as females, only five of fourteen subjects were firmly committed to a female identity. Eight rejected their reassignment—some of them spontaneously and some after their parents revealed to them the truth about their condition—and embraced a male identity. (The remaining subject

was uncooperative with the authors of the study.) The two subjects who did not undergo surgery and were raised as males
appeared secure as males.

The lesson from this rare glimpse of the aftermath of sex
reassignment is clear: We cannot yet predict with precision the
outcome of sex reassignment at birth and so we should be more
humble in our attempts to manipulate sexual development. This
humility stands in stark contrast to the certitude of John Money
concerning the plasticity of human sexuality at birth. Paul
McHugh, citing the study of cloacal exstrophy and drawing
upon his other experiences as a psychiatrist, advocates a measured and rational approach to children born with ambiguous
genitalia. Specifically, he urges surgeons to correct immediately
any major problems that threaten the health of the infant,

> but to postpone any decision about sexual identity until much
> later, while raising the child according to its genetic sex....
> Settling on what to do about it should await maturation and
> the child's appreciation of his or her own identity.... This
> effort must continue to the point where the child can see the
> problem of a life role more clearly as a sexually differentiated
> individual emerges from within. Then as the young person
> gains a sense of responsibility for the result, he or she can be
> helped through any surgical constructions that are desired.
> Genuine informed consent derives only from the person who
> is going to live with the outcome and cannot rest upon the
> decisions of others who believe they "know best."[31]

This perspective, coming from a psychiatrist, meshes with the emerging views of intersex individuals themselves.[32] All agree that surgery may eventually be an option for some, but current procedures can produce results that are so traumatic and regrettable that many now feel that the decision to go forward with surgery cannot be made on behalf of someone else.

In the meantime, alternative, non-surgical treatments are emerging that might prove more promising for some. For example, in a study of males with micropenises that resulted from deficient prenatal levels of testosterone, a regimen of testosterone injections—in infancy or childhood—appeared effective for lengthening the penis.[33] The earlier the treatment, the more effective it was, but all the subjects in the study showed improvement. Many intersex conditions will not respond so readily to so simple a treatment, but the point of greater significance is that modern medicine is beginning to embrace a less destructive and more nuanced approach to the management of intersexuality. Perhaps, one can hope, the days of monster ethics are finally drawing to a close.

OBJECTS OF AFFECTION

John Money seems to have tragically underestimated the significance of prenatal factors in the development of our gender identity, but what all those factors are is far from certain. Nor is the notion of prenatal establishment of a fixed identity at birth the only alternative to Money's views. A biologically realistic perspective would instead consider sexual identity as a meandering, unfolding path that begins early in embryonic development and continues after birth and well into the early years of life. As the embryo and infant moves along this path, it acquires sex characteristics that distinguish it as a male or female. Sex chromosomes influence the "choice" of testes or ovaries; if testes develop, then the release of testosterone and the presence of testosterone receptors influence the development of internal and external genitalia. Thus, as development proceeds, the male and female paths typically diverge and head off in distinct directions.

But the vagaries of development can produce alternative paths and short cuts that effectively break down our standard conceptions of male and female. XY males with testes but without testosterone receptors resemble females more than males; XX females with excessive androgens exhibit masculinized external genitalia; XY males that cannot convert testosterone to dihydrotestosterone (like Callie Stephanides) exhibit ambiguous genitalia at birth. The list of possibilities is long.

At the time of birth, the male and female paths may be more or less distinct, but it appears that the destinations are not yet fully determined. It is not possible to examine a newborn baby and read its future. Perhaps relevant information is hidden within the fine details of hormone receptors located within the brain—but we do not yet have the capacity to extract that information. Perhaps timing matters, such that the path to sexual identity for a premature infant, for example, is not as well defined at birth as the path for a full-term infant.

Regardless, although male and female human newborns are traveling on different paths to sexual identity, they must still make the journey. The destination is not fixed, and it does not exist anywhere within the child. Even the path does not yet exist. It is rolled out, like a red carpet before royalty, as the child interacts with its world through developmental time.

This development continues long after birth. Psychologists have found that children begin to distinguish boys from girls at around eighteen months of age, but that some do not make this distinction for ten more months.[34] Those who can make this distinction earlier exhibit more pronounced sex-typical behavior.

Moreover, parents influence this developmental timing through their own behavior as they direct their child's attention toward toys and activities that are "appropriate" to the child's gender. In this way, the child's "gender schema" is constructed through daily interactions with parents, other children, and objects in the environment. Moreover, those interactions cannot be divorced from the anatomical, physiological, and behavioral features of the child that emerge during prenatal and postnatal life.

Our sexual attractions may also emerge in the same way that every other biological feature emerges—in developmental time. According to Daryl Bem's "Exotic-Becomes-Erotic" theory,[35] sexual orientation emerges developmentally through an interaction of biological, experiential, and sociocultural factors. Rather than view sexual orientation as a genetically determined trait, Bem views genes, hormones, and other biological factors as shaping the child's level of activity and style of play, which in turn shape that child's interactions with other children. Ultimately, Bem suggests, individuals "become erotically attracted to a class of individuals from whom they felt different during childhood."[36]

Within the broader framework of sexual development, such epigenetic approaches to sexual orientation make sense. If our gonads are not genetically determined, if our genitalia are not genetically determined, and if our style of play is not genetically determined, then why would we believe that our erotic attractions would be? This does not mean, of course, that we *choose* these attractions, any more than we choose our genes or our gonads.

Still, the irresistible pull of "genetic explanations" remains, and so we can expect that many will continue to cling to the notion of, say, a 'gay gene' as an effective refutation of the argument that homosexuality is a "lifestyle choice." But we should strive to separate politically expedient argument from biological reality. At best, even the vaunted X and Y chromosomes can only help to get the developmental ball rolling; they do not determine outcomes. Thus, the expectation of some that our sexual attractions *must* be encoded in DNA—and the associated view that such encoding is central to any claim about the *biological basis* of sexual attraction—is both unfortunate and false.[37]

The intertwining of *biological* with *genetic* emerged under the influence of the Modern Synthesis, which focused on evolution as a process of genetic change. Over time, the argument emerged that any biologically important behavior must have a genetic basis if it is to be reliably inherited and expressed across generations. Such a genetic foundation, according to this once unassailable perspective, provides a necessary buffer against an unreliable and often turbulent environment.

It was on the basis of such arguments that Ernst Mayr, Konrad Lorenz, and other influential biologists based their views on the evolution of instincts.[38] According to them, instincts are biologically essential behaviors that are "written in the genes." For example, although the offspring of some bird species are incubated and hatched in the nests of other species—such so-called *brood parasites* include cowbirds and cuckoos—they nonetheless manage to locate and mate with members of their own species. Similarly, even when male zebra finches are raised in the nests of Bengalese finches, they mature into males that prefer to mate with female zebra finches.

These vivid examples of *sexual imprinting*—that is, the development of a sexual preference for members of one's own species—were once explained by invoking the existence of what Mayr called *closed genetic programs*.[39] Such programs, according to Mayr, ensured the timely expression of adaptive, instinctive behaviors regardless of the animal's early experiences.

To appreciate why these and other such notions have crumbled over the last several decades, one need only appreciate this simple insight: Juvenile birds learn to recognize members of their own species because the adult members of those species recognize *them*, approach *them*, and engage *them* in behavioral interactions.[40] These interactions provide the developmental experience upon which sexual imprinting is built. Indeed, through experimental manipulation of these early experiences, birds can be induced to mate with members of the "wrong" species. What has evolved, then, is not a genetic program self-contained within each individual bird, but a biological *system*—an intergenerational fabric woven by adults and juveniles engaged in complex behavioral interactions. Moreover, as is now well known, the entire *system* or *niche* is inherited—in the broadest sense of that term—just as surely and reliably as a beak or a wing.[41]

If birds must learn to recognize members of their own species for the purpose of mating—a capability whose evolutionary significance is not in dispute—then we should finally be willing to put to rest the notion that similar capabilities in humans are "genetically determined." We are the sum of many parts and processes—cascading effects of genes and gonads and the secreted hormones that stimulate genital growth and modify the functioning of our brains. All these factors conspire to construct the person we see in the mirror. But so do our numerous interactions with others that define how we are perceived and, ultimately, who we are as individuals embedded in a family, culture, and time. We may see only one individual in the mirror, but we are never truly alone.

HAVING IT BOTH WAYS

As the example of sexual imprinting illustrates, and as we have seen many times in this book, non-human animals can help us to understand complex issues that are often more difficult to frame when considering humans. We can, for one thing, more easily set aside political and social sensitivities about sexual preference when we are discussing, for example, a cowbird. What, then, can we learn from others in the animal kingdom about the no-less-complicated topic of sexual development, whether anomalous to a given species or so diversely evolved as to confound our neat gender categories?

Consider these examples of intersexuality in animals found in the wild:

- A beluga whale adrift in the St. Lawrence Estuary in Quebec, Canada, was found to have two testicles, two ovaries, and two complete sets of male and female internal sex organs.[42]
- An "antlered doe," shot in South Carolina, possessed testes and male internal organs along with female external genitalia, including clitoris and vagina.[43]
- A growing population of egg-bearing male smallmouth bass has been reported in the Potomac River.[44]
- Polar bears in Svalbard, an archipelago in the Arctic Ocean in the northernmost part of Norway, have been found that sport a vaginal opening with enlarged penis-like clitoris, complete with baculum.[45]

- And another polar bear, a male with hypospadias, had the "good fortune" of finding surgeons willing to amputate his penis, extract his testes, and create an opening between his legs for the passage of urine.[46]

There are also numerous reports of intersexuality in domesticated animals, including sheep, mink, pigs, dogs, cats, rats, horses, and goats.[47]

Such freakish cases of ambiguous sexuality are typically dismissed as evolutionarily inconsequential. After all, intersexes are typically unfit for reproduction—that is, they are unable to attract mates, copulate, or bear offspring—and therefore are considered outside the narrative of evolution. But they are *not* outside the narrative—they are in fact an integral part of it. A recent television interview with a constitutional law professor inadvertently clarified this point for me. The professor was emphasizing that the U.S. Congress has the responsibility to pass laws that are clearly written so that those laws are not open to multiple interpretations. Ambiguity, he said, breeds mischief.

Well, ambiguous biological *mechanisms* breed evolutionary mischief, and sexual ambiguity is one of the results. In mammals, it begins with the sex chromosomes and the uncertainty with which they determine the identity of a gonad. It continues with the gonad, which is initially called an *ovotestis* to highlight its capacity to develop into an ovary or a testis. It continues with the hormones secreted by the gonads and other organs, the complex and easily altered processes that convert one hormone into another, and the receptors that must also exist for the hormones to have their effects. It continues with the internal genitalia, which exist initially in both the male and female forms until one is retained and the other disappears. It continues with the external genitalia, characterized by a clitoris, which is capable of becoming an elongated penis, and by labia, which are capable of fusing into a scrotum. And it continues with a brain whose developmental experiences with testosterone, estrogen, and other hormones shape the kind of brain it will be.

This is not a system *capable* of ambiguity. This is a system *guaranteed* to produce it.

Put another way, the developing fetus is *bipotential*, by which we mean that it has the capacity to develop in a male or female direction. The ovotestis is *bipotential* because it comprises ovarian and testicular tissue. The internal genitalia are *bipotential* because both male (Wolffian) and female (Mullerian) systems are initially produced before one typically disappears. The external genitalia are *bipotential* because the male forms are mere elaborations of the female forms.[48] Finally, the brain is *bipotential* because it can support either male or female behavior.

For each of these organs and systems, hormones and their receptors play a major role in tilting development in the male or female direction. For example, when an ovotestis becomes a testis, it secretes testosterone and thereby stimulates retention of the Wolffian system. A second hormone of testicular origin, called *Mullerian inhibiting factor* (MIF), causes the

Mullerian system to disappear. If one of these hormones, for whatever reason, cannot act properly, the result can be either the retention of two sets of internal genitalia—both male and female—or the loss of both. The latter case is seen in humans with Androgen Insensitivity Syndrome, the condition in which individuals with XY chromosomes and testes nonetheless mature into feminine adults. With their normal testes, individuals with this condition produce testosterone as well as MIF. But because these individuals do not possess testosterone receptors, their testosterone cannot have any effect. As a result, the Wolffian system disappears. Moreover, because MIF functions just fine, the Mullerian system also disappears. The result is a woman with XY chromosomes, testes, testosterone, feminized external genitalia, and nonexistent internal genitalia. Such is the ambiguous nature of this developmental system.

One of the oldest observations of ambiguous sexual development concerns the so-called free-martin, described in 1779 by John Hunter, a Scottish scientist and surgeon, in this way:

> It is a known fact, and, I believe, is understood to be universal, that when a cow brings forth two calves, and that one of them is a bull-calf, and the other a cow to appearance, the cow-calf is unfit for propagation; but the bull-calf becomes a very proper bull. They [that is, the cows] are known not to breed: they do not even shew the least inclination for the bull, nor does the bull ever take the least notice of them. This cow-calf is called in this country a *free martin*....[49]

As Hunter indicated, the external appearance of the cow was female, so the masculine appearance of the internal genitalia—and the notable absence of a female system—created confusion. Most scientists saw the free-martin as an undermasculinized male until 1916, when Frank Lillie, a zoologist at the University of Chicago, brought statistical and embryological evidence to bear on the problem. Lillie suggested that hormones released by the male co-twin passed in the blood through a shared circulatory connection to the female embryo and, as a result, both defeminized her (by causing the regression of the Mullerian system) and masculinized her (by stimulating retention of the Wolffian system).

As we now know, the hormones performing these actions in the free-martin are MIF and testosterone, respectively, but Lillie did not know this when he proposed his theory. Rather, his enduring contribution was to help establish the notion itself that the sexual organs in mammals develop from a bipotential state under the influence of hormones.[50] His contemporaneous assessment that with the free-martin "nature has performed an experiment of surpassing interest"[51] was prescient, as his insights were to reshape research on sexual differentiation and hormone action for the rest of the twentieth century. Indeed, our understanding that sexual differentiation depends to a large extent on the action of hormones, such as testosterone and estradiol, rests on the groundwork laid by Lillie's research.

We are now comfortable with calling androgens like testosterone a "male" hormone and estrogens like estradiol a "female" hormone. But it cannot be overemphasized that these two hormones, which exist in both sexes, are identical except for the smallest of molecular changes in their chemical structure. Those changes are brought about by proteins called *enzymes*. In the case of androgens and estrogens, the process begins with a single precursor molecule, cholesterol, and with its conversion to pregnenolone, which can be converted to progesterone, then to 17α-hydroxyprogesterone, then to testosterone, and then to estradiol. Each conversion requires its own enzyme, and each enzyme is regulated to control a particular hormone-dependent developmental process.

For example, testosterone is normally produced in the placenta (part of this hormone regulatory system) from precursor molecules that arrive from the adrenal glands (another source of steroid hormones) of both the fetus *and* the mother. The aromatase enzyme within the placenta converts the testosterone to estradiol. But if this enzyme is blocked, testosterone levels will swell; the effect is like damming one river outlet, causing the river's other outlets to overflow. One recently discovered intersex condition in females is caused by this very blockage, resulting in an aromatase enzyme deficiency that has been attributed to a genetic mutation.[52] In the absence of aromatase, levels of testosterone and the related androgen, dihydrotestosterone, are elevated in the fetus, resulting in the masculinization of her external genitalia.

Androgen Insensitivity Syndrome, intrauterine exposure to hormones as in free martins, and aromatase enzyme deficiency are only three of the phenomena that can deviate sexual development. Many others are produced when various hormones, enzymes, or receptors are altered in one way or another, at one time or another. Many of these conditions can be traced to influences communicated to the fetus from the mother, as when the mother has a hormone-producing tumor or is exposed to environmental chemicals that mimic the effects of natural hormones. Other conditions can be traced to genetic mutations in the child that alter the activity of hormones, enzymes, or receptors.

In general, most of us think of these effects as downstream from the ultimate source of sex determination—the sex chromosomes. Indeed, in most species, sex chromosomes do help to determine whether an ovotestis develops into an ovary or a testis.[53] They just do not do it alone; hormones and other factors also come into play to advance, and sometimes deviate, the sex-determining process.

But what about those species—crocodiles and various species of turtles, lizards, and fish—that do not have sex chromosomes? These are species with sexes that are as identifiable as those in any mammalian species, yet their sexual determination is hardly "written in the genes" because they simply do not have the sex genes to be written on. In fact, these genetically identical embryos are equally capable of becoming males or females, because what determines sex in these species is the *temperature* at which the embryos are incubated.[54] This phenomenon is known as *temperature-dependent sex determination* (TSD) to distinguish it from the more familiar genetic sex determination (GSD).

In those species with TSD, eggs are laid, often buried in sand. During the middle period of gestation, a temperature-sensitive "decision" is made, tilting each embryo's ovotestis toward ovary or testis. In some species, clutches of eggs predominated by males are produced at high incubation temperatures, and clutches predominated by females are produced at low incubation temperatures; the reverse is true in other species. In still others, males are produced at moderate incubation temperatures and females are produced at the extremes. Regardless of the exact pattern, at female-producing temperatures the aromatase enzyme is activated, thereby facilitating the conversion of testosterone to estradiol and, ultimately, the formation of an ovary. At male-producing temperatures, 5-alpha-reductase is activated, thereby facilitating the conversion of testosterone to dihydrotesterone and, ultimately, the formation of a testis. In addition, estrogen and androgen receptors are produced at female- and male-producing temperatures, respectively.[55] Intersex offspring are rarely produced.

TSD is not a primitive process that was replaced by GSD in "higher" species. Rather, these two modes of sex determination can be found in closely related species, thus suggesting that the two processes, once considered dramatically different from one another, are actually a single process comprising a shared network of epigenetic events. What differs, it seems, is simply those factors that propel the ovotestis toward an ovary or a testis.[56]

Recent work is pushing this perspective even further. Scientists working with the central bearded dragon, a lizard indigenous to Australia, recently showed that the possession of sex chromosomes does not preclude the action of temperature-determining sex effects.[57] By incubating genotypic male lizards at high incubation temperatures (that is, between thirty-four and thirty-seven degrees Celcius), the researchers demonstrated an increasing tendency for females to be produced. Lower incubation temperatures produced hatchlings consistent with their genotype. The current theory is that the high incubation temperatures inactivate a gene product associated with the male's sex-specific genotype, lowering this product below a threshold and thus preventing formation of testes. In this way, temperature trumps the genes. These lizards have it both ways.

Thus, it now seems most accurate to view the two "modes" of sex determination as occupying opposite ends of a continuum, with the central bearded dragon being the first intermediate form to be discovered along that continuum. Accordingly, temperature and sex chromosomes are *interchangeable* influences on sex determination, each capable of affecting a system that, in different species, shares most of the same components. Thus, for example, the evolutionary transition from GSD to TSD would require only that "sex determination be 'captured' by an environmental influence such as temperature at any number of points in the pathway leading to male or female differentiation."[58]

TSD is a remarkable phenomenon, powerfully illustrating how animals are best perceived as systems with developmental histories rather than as individuals with intrinsic essences. Males and females *develop* into dramatically different individuals—they do not start out that way.

Next we will meet animals whose very existence demolishes any remaining sense we might have that males and females occupy clear and distinct regions of the biological universe. As we will see, the oddity of an intersex individual fades away as we encounter entire species that have evolved intersexual lives.

THE LONG AND THE SHORT OF IT

In the previous chapter, we saw that as the ancestors of whales traveled the evolutionary road from land to sea, their limbs were extensively modified. As a result of these and other modifications, whales have adapted to life in the water. But in contrast to those species that have evolved their limbless lives over many millennia, individual "freaks" like Johnny Eck must adjust to their limblessness within a single lifetime. Still, as we now know, limbless species and individuals share a deep connection: their limblessness is produced through manipulations of one and the same limb-development system.

As with limbs, so with sex. The free-martin, an individual intersex oddity, arises when the hormones produced by a male embryo exert developmental effects on a female twin. The adult free-martin is infertile and therefore an evolutionary cul-de-sac.

But the hormonal influence of male fetuses on female fetuses does not always result in infertility. In those species that produce litters comprising numerous offspring—such as pigs, mice, and gerbils—female fetuses sandwiched between two males are reliably exposed to sufficient testosterone to alter their anatomy, physiology, and behavior.[59] Importantly, female fetuses masculinized by this early hormone exposure are not only able to reproduce, but their behavior, including increased aggression and territoriality, can even be advantageous for survival and reproduction. In this way, prenatal hormonal experiences enhance female variability, a potentially powerful reproductive feature for species that give birth to multiple offspring in often unpredictable and harsh environments.

Prenatal exposure to high levels of androgens, you may recall, also describes congenital adrenal hyperplasia (CAH), the most common intersex condition in human females. CAH is characterized by a range of outcomes—from a slightly enlarged clitoris in an otherwise typical female infant to a penis-like appendage that can fool the parents for some time into raising their child as a boy.

Interestingly, a similar range of intersex conditions was reported in the 1980s in several wild brown and black bears.[60] In one adult female brown bear, the clitoris resembled a small penis with a baculum. But because the urethra failed to traverse the penis and opened instead near the vagina, this female penis was hypospadic.

Even more extreme, an adult female black bear was found with "normal ovaries, oviducts, uterus and uterine horns, cervix, and anterior vagina, but the anterior end of the vagina developed internally into a urethra and emerged externally as a penis-like structure with urethra and baculum."[61] The "penis" itself was impressively sized—about three-quarters the size of the average male black bear's—and properly located. This bear also possessed a male-like urethra. But at the time of her death this female was nursing two cubs. Moreover, she had placental scars in her uterus, signs that the cubs were indeed hers. What this means, of course, is that this intersex black bear must have urinated, copulated, and given birth through her penis-like clitoris.

No evidence exists that the intersex condition of these bears resulted from CAH or a related condition. Of the many possible causes, natural or contaminated foods may have exerted masculinizing effects on these bears during their gestation. Regardless, the fact that a female bear could manage the difficulties of copulating with her penis-like clitoris (to say nothing of the challenge this must have posed for the male) and giving birth through such a narrow canal is a marvel of behavioral and anatomical flexibility. This was one remarkable bear—but still only a single freakish occurrence.

Spotted hyenas are different. In this African carnivore, every female has a penis-like clitoris, every female lacks a vagina, and every female has fused labia that form a scrotum—and this scrotum is filled with fat, giving the impression of testes contained within. Also, every female has internal sex organs that are typically and functionally female.

The behavior of female hyenas matches their external appearance. Similar to males, the female's enlarged clitoris can be erected and displayed during nonsexual interactions referred to as *meeting ceremonies*. Moreover, in a species notable for its aggression, the female is especially so. Thus, the masculinization of her form is complemented by the masculinization of her brain, producing a fully realized and uniquely modified intersex mammalian species.

The female hyena's clitoris, like that of the anomalously intersexed black bear, must serve three critical functions: urination, copulation, and birthing. Urination is straightforward. Copulation is trickier, as the sexually engaged male hyena must flip his penis into the tip of the elongated clitoris, a task made somewhat easier by his longer, narrower shaft and angled tip.

 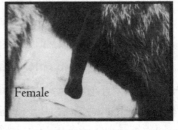

The semierect penis of a male spotted hyena (left) and
the erect clitoris of a female (right).

Birthing, however, may present the greatest challenge posed by the female's elongated clitoris. The challenge begins when the fetus's umbilical connection—and thus its supply of air—is cut off early in the birth process, thus requiring the fetus to make the journey through its mother's unusually long and narrow birth canal without a source of oxygen. For first-time mothers, the birth process also requires internal tearing to make way for the fetus. These challenges translate into a considerable cost—as many as sixty percent of first births culminate in a dead newborn. Subsequent births are much more successful.

The peculiar sexual features of female spotted hyenas have only compounded the many unfavorable characterizations that have been heaped upon the species as a whole by indigenous Africans and early European writers alike. Hyenas have been called scavengers, cowards, gluttons, and idiots of comic proportions.[62] To say the least, hyenas have never been lionized. Moreover, the female's sexual idiosyncrasies have been distorted and misunderstood: some have claimed that hyenas switch between sexes, while others have believed them to be hermaphrodites.

Modern scientists, however, do not refer to female spotted hyenas as hermaphrodites because that designation is typically reserved for those individuals that simultaneously exhibit features of both sexes. And although female hyenas do indeed exhibit many features that we normally associate with the typical male mammal—including the males of other hyena species—they do not simultaneously exhibit features of both *spotted hyena* sexes. That is, modern female spotted hyenas are clearly distinct from modern male spotted hyenas. On the other hand, when the spotted hyena was first emerging as a distinct species, those first females with penis-like clitorises would have appeared as intersexes in relation to all of the other *typically female* hyenas (just like human females with CAH or the female black bear discussed earlier). But as time passed, the intersex females displaced the "normals" and became the standard-bearers of this now-freakish species.

The search for the causes of the female hyena's enlarged clitoris made clear just how marvelous these animals are.[63] This search began with the reasonable expectation that female hyenas, like the females of numerous other mammals that had been studied, might be masculinized through exposure to high androgen levels at some point during the fetus's 110-day gestation. Such an explanation would fit with the classic model of sexual differentiation, accepted since the 1950s, according to which testicular androgens play a preeminent role in the formation of sex differences.

With hyenas, however, the simple application of this classic model did not seem to work. Thirty days after fertilization, female embryos already exhibit the developmental precursors of an enlarged clitoris, as well as the scrotum. Yet neither their ovaries nor their adrenals are able to produce the androgens that one might think would be necessary to masculinize the external genitalia. Of course, androgens could be produced by another fetal organ or transferred to the fetus from the mother. But even if a female embryo as early as twelve days post-fertilization is treated with a drug that blocks the activity of androgens—a procedure that blocks the effects of androgens whatever their source—it still develops an enlarged clitoris and scrotum. Clearly, some non-traditional mechanism—hormonal or non-hormonal—is at play here.

Further clues to the organization of the hyena system come from examining the timing and mechanisms of sex differentiation. The fully developed female clitoris, though uniquely large, is not identical in size, width, or shape to the male penis, but through the first month of gestation, male and female phalluses are indistinguishable. It is testosterone, produced by the testes after the first month, that causes the internal and external features of the male phallus to diverge from those of the female.

In this way, testicular androgens are responsible for the differentiation of the phallus *beyond* the "prematurely" enlarged and undifferentiated state. Exposing male fetuses to androgen-blocking agents brings this observation home. Whereas such agents administered to female fetuses do not prevent the formation of an enlarged clitoris,[64] in males the effect is profound: They cause the male's phallus to look like a typical female's. In other words, males treated in this way are *feminized*, now exhibiting *female-typical penis-like clitorises*. But note: "feminized" here refers to the hyena form of feminization, and these "penis-like clitorises" are penis-like only in comparison to the penises of other male mammals, not male hyenas.

It would seem that an enlarged clitoris and fused labia are the foundation upon which the sex differences of hyenas, unlike other mammals, are built. There is still some mystery as to the underlying masculinizing mechanism. In particular, the search is on for an experimental treatment that will produce a female hyena possessing a standard diminutive clitoris—that is, the kind of clitoris found in nearly every other female mammal.[65] As this search continues, we can be confident that evolution discovered a novel way of manipulating development to alter the undifferentiated state of the embryonic spotted hyena, in effect *raising the platform* upon which further differentiation rests.

It is this raised platform that makes spotted hyenas so special. But we should not get hung up on this aspect of hyenas. More important is the particularly compelling testimony they give to the notion that evolutionary novelties, whether mundane or marvelous, arise through modifications of development. Some freaks live lonely lives. Others hunt in packs. Both demonstrate the intimate ties that bind individual anomalies and evolved characteristics.

CONSIDER THE ALTERNATIVES

Female hyenas may be odd, but they only begin to stretch the bounds of sexual disorientation. After all, female hyenas are unquestionably female. But what we observe throughout the rest of the animal world is a diverse and overwhelming collection of unisexuals, intersexuals, and multisexuals that should

sap any remaining strength we might have to sustain an impreg-
nable wall of separation between the sexes.

When Cal Stephanides, the protagonist of *Middlesex*, noted
that his "change from girl to boy was far less dramatic than the
distance anybody travels from infancy to adulthood," he could
have been describing the experience of many fish. In fact, fish
may be the champions of sexual flexibility.[66] For some species,
switching sex is about as dramatic as changing clothes.

Imagine the benefits to your reproductive output if you could
impregnate every female in the neighborhood and then turn
around and be impregnated yourself. Such is the promiscuous
life of the tobaccofish. This fish—a Caribbean seabass related
to the grouper—is a prototypical example of a *simultaneous* hermaphrodite, an animal that is able to function as male or female
from one moment to the next.[67]

When a pair of tobaccofish begins a courtship and mating sequence, one of the fish initiates the sequence by playing
the female role. Beginning near the sea floor, "she" approaches
another fish, called the follower, and may quiver in front of and
bump up against "him." If the follower is responsive, "he" will
bump into the initiator and bumping will continue as the pair
rise toward the water surface.

Mating begins when the initiator turns toward the follower
and the two fish rise quickly together toward the surface in a
"mating rush" that culminates in the explosive release of eggs
into the water by the initiator and of sperm by the follower. The
eggs, now fertilized, float away.

But our two tobaccofish are not done with each other. Rather, they usually return together to the sea floor where, within minutes, the follower switches to the role of initiator. The sequence of courtship and mating is then repeated, but with roles reversed. Such "egg trading," as it is called, can occur many times each day, with each fish playing the role of male or female with equal frequency. This tit-for-tat system of reproductive cooperation distributes benefits evenly across a group of cooperative individuals. But it only works, and endures, when fish play fair—and for now, for the sake of reproduction, tobaccofish play fair.

Clearly, the tobaccofish illustrates an extreme form of plasticity—one that is expressed almost without delay. Other forms of sexual plasticity in fish can take several days to complete. In particular, *sequential* hermaphrodites function as one sex, then switch to the other sex when circumstances dictate. For example, many tropical fish that live in coral reefs exhibit a *harem* style of reproduction in which a single, dominant male controls the reproduction of a group of females. (The opposite pattern, in which a dominant female controls a group of males also exists, but is less common.)

But a funny thing happens when the dominant male dies or disappears: One of the remaining females—usually the dominant one—transforms into a male and consequently takes over the harem. With this transformation comes a dramatic increase in the ability to produce offspring, as this newly minted male now controls the reproduction of all the females in the harem.

The fact that the dominant female is typically the one to transform into a male reflects an underlying principle that appears to guide sexual decisions in sequential hermaphrodites. That principle is this: "If an individual can reproduce more effectively as one sex when small or young and as the other sex when larger or older, it should change sex at some point in its life history."[68] Obviously, switching is not an option for humans because our developmental systems do not abide such sexual flexibility. But for those fish and other animals that are flexible, scientists have been able to identify a variety of developmental, ecological, and social factors that have shaped the evolution of these sexual decisions.[69]

The bluehead wrasse, a small Caribbean coral-reef fish, illustrates especially well the social modulation of sexual differentiation while also introducing still another dimension of sexual complexity. These fish have one kind of female but *two* alternative forms of males that are best suited to two different social conditions.

Imagine that you are a juvenile bluehead wrasse living on a sparsely populated coral reef.[70] In such conditions, the few large and brightly colored males on the reef defend a few breeding sites where females come to mate with them. In such a highly competitive situation, with a few dominant males monopolizing the situation, your best bet is to differentiate into a female, grow larger, and switch to a male if and when the opportunity presents itself. Certainly there is no point in switching if you cannot successfully compete for a breeding site. So, you bide your time and wait for one of the large, colorful males to die or leave. When one does, you make the switch and take his dominant position as quickly as possible—a maneuver that lets you grab the open spot and also suppress any other females on the verge of switching.

Completing a sex change can take several days while the ovaries develop testicular functionality. So the interim period can be tenuous to the extent that other females may also make the switch, resulting in an overabundance of newly switched males of which only one would be able to take and hold a breeding site. Thus, because suppressing other females is such an important part of this process, and because suppression is caused by the presence of nearby fish that behave like males, sex-switching fish begin to behave like males within only minutes or hours of the loss of a dominant male. This rapid behavioral transition is accomplished by the fish's brain without any contribution from the gonads.[71] Once male-like behavior is in place, the fish can afford to wait for the gonads to "catch up."

Female bluehead wrasses that switch sexes become what are called *secondary males*. Because the testes of these males were once ovaries, they retain this ovarian history in the testicular tissue itself. In contrast, *primary males* develop testes during early development without passing through a female stage, thus distinguishing them anatomically from secondary males. These primary males—which are small and lack the bright coloration of secondary males—predominate on reefs with large populations of fish. Such conditions make it impractical to defend a breeding site as a means of attracting females. Instead, these males maximize their reproductive success by spawning in large groups.

Thus, with bluehead wrasses, we have two separate paths to becoming a male: a *primary* path that is preferred when the local population is large, and a *secondary* path that is preferred when the local population is small. This means that the nervous system of a juvenile fish must accurately assess local fish density before committing to one of these paths. Similarly, the nervous system of an adult female must assess social information—that is, the presence of a dominant male—before triggering a sex switch. What has evolved in this species, then, is a developmental system that can flexibly respond throughout the lifespan to local environmental conditions in order to compete more effectively for mates.[72]

Sex switching distorts our definition of sex, but some fish go even further and challenge our definition of individuality. For example, some species of deep-sea anglerfish, swimming in pitch-blackness 300 meters beneath the ocean surface, have evolved a unique form of reproduction that begins with females that can be as much as one-half-million times larger than males.[73] The male's evolved response to this lopsided situation is written on his face: He is equipped with oversized eyes and nostrils—for locating a female—and a pincer-equipped mouth to dig into her flesh and hold on. Once attached, his individual identity fades away as their bodies and bloodstreams fuse. At this point, the male becomes little more than a decaying container of testicular tissue. The female provides oxygen and nourishment to him—and to other males that can be similarly attached—in exchange for sperm. This mode of reproduction has been termed *sexual parasitism,* but one can also see why these fish, when fused, might reasonably be called *hermaphrodites*— intersexes comprising internal ovaries and outboard testicles.

A female anglerfish (left), 1.8 inches long, with a parasitic male attached to its back. The male (in close-up at right), only a quarter inch long, is one of the smallest known vertebrates.

If male anglerfishes are diminished, male Amazon mollies must be counted among the disappeared. Named after the mythical all-female nation, the Amazon molly is a unisexual species found in Texas and northern Mexico. In the absence of males, reproduction in this species must be accomplished asexually—that is, mollies are clones of their mothers. But this does not mean that mollies do not mate with males. They do, but with the males of other, closely related species. During mating, these males transfer sperm to the female,[74] but these sperm serve only as a chemical trigger for growth of the female's embryos. There is no genetic transfer. Odd as it may seem, these mollies evolved beyond a need for male DNA, but not for male sperm.[75]

Leaving the fish world for a moment, perhaps the most famous unisexual vertebrate is the whiptail lizard. It has evolved the capacity to reproduce asexually and, unlike the Amazon molly, it has no need for sperm. But again, the absence of male whiptail lizards does not mean that the female's ties to her sexual evolutionary past are completely lost. On the contrary, during a mating sequence, one female will mount another female—a *pseudosexual* behavior that resembles the typical male mounting behavior observed in closely related species (but again, without the transfer of sperm). Importantly, females that are mounted by other females produce more young than those that are not. Later, during the same breeding season, a female that was previously mounted will now do the mounting, and vice versa, in a pattern that resembles the more rapid role switching of tobaccofish. This capacity to exhibit female- and male-like behavior in the same animal is no mean feat, and likely required an evolutionary restructuring of the development of this species to allow for more flexible interactions among gonads, hormones, and brain.[76] Thus, contrary to the expectations of some human females, life without males is not necessarily less complicated.

———

SEXUAL DIFFERENTIATION TYPICALLY refers to the development of differences *between* sexes. But sexual differentiation can refer to more than just between-sex differences, and we caught a glimpse of this with the bluehead wrasse's primary and secondary paths to maleness. However, these fish only begin to capture the within-sex diversity that exists in nature. Indeed, in some species, within-sex variation can be as extreme as between-sex variation. It is time to meet the multisexuals.

In fact, you have already met one group of multisexuals: the dung beetles from the previous chapter. There you saw that male beetles come in two varieties—big-bodied with horns and small-bodied without—and that the horns are used as weapons to guard a tunnel—and the reproductive female within—against all other horned challengers. But what about those small, hornless males? Must they resign themselves to sulking at the end of a beetle bar, juiced on fermented dung, horny but hornless, lamenting their fate as losers in a cruel, cruel world? Well, not quite.

If a hornless male attempted to fight his way into a guarded tunnel, the large, horned male would eject him like a bouncer at a trendy nightclub. So, rather than fight a losing battle, a hornless male will try to sneak past the horned male at the tunnel entrance or dig a second tunnel and bypass the horned male entirely.[77] Once safely inside, this *sneaker* will surreptitiously mate with the female and then escape without detection. His small size and hornlessness are well suited to his stealthy lifestyle—he is agile and quick of foot, and there is no risk of horns scraping along the tunnel and producing vibrations that might announce his presence to the guarding male.

So there is a suitable match between a male beetle's equipment and occupation. But we must remember that, like environmental temperature determining sex in a turtle, *nutrition* is the key factor when it comes to a male beetle's body size and, therefore, his horns (or lack thereof). Moreover, anatomy and behavior are modified together—horned males guard and hornless males sneak. Although it is not clear how this linkage between anatomy and behavior evolved, develops, or is learned, it is clear that nutrition is a critical early factor in determining the kind of male a beetle will be.

Sneaking in male beetles is an example of an *alternative reproductive strategy*. Such a strategy typically entails some developmental "decision" that diverts an animal down a path toward more successful reproduction than if another path had been taken.[78] The key consideration here is the matching of the conditions of early development—for example, nutritional status—with adult form and behavior. A nutritionally disadvantaged larva that grows into a small, hornless male would not be an effective guard, and a well-fed larva that grows into a large, horned male would not be a successful sneaker.

A similar relationship between body size and reproductive strategy can be found in numerous other species, especially fish. For those species of fish in which the female releases her eggs directly into the water, any male that is able to release sperm in the vicinity of the eggs can be a father. When more than one male gains access to the eggs and releases sperm, the sperm themselves then compete to successfully fertilize the available eggs. This highly competitive environment, coupled with extreme variability in the growth rates of fishes, seems to have spurred the evolution of alternative reproductive strategies in these animals.

In part because of their importance as a source of human food, salmon are among the most widely studied animals that exhibit such strategies. Coho salmon, which populate the waters between California and Alaska, begin their lives as hatchlings in freshwater streams. At one year of age, both males and females migrate to the open seas. Those that survive will eventually return to their natal stream to breed and die.[79]

When males return from the ocean, they do so as members of two distinct groups. First, there are the large, brightly colored males that possess equipment suitable for fighting, including a hooked jawbone with sharp teeth and protective cartilage along the back. These *hooknose* males will spend 18 months in the ocean before reaching sexual maturity and returning to their stream to mate. The second group comprises small, cryptically colored males, called *jacks,* which mature after only six months in the ocean before reentering the stream.[80]

Living together in the same stream, the competition between hooknoses and jacks commences. The large and well-equipped hooknoses fight for the position of honor closest to a female on the verge of releasing eggs. The victorious hooknose swims ahead of his lesser competitors like a movie star with his entourage. Meanwhile, away from all this action, the jacks skulk and prowl, awaiting their opportunity to sneak a fertilization.

Although salmon return from the ocean as hooknoses and jacks, their fates are set during the year between hatching and their departure for the ocean. In fact, soon after hatching, differences in social status and body size can predict future rates of growth, age at sexual maturity, and, ultimately, development as hooknoses or jacks.[81] In general, jacks exhibit rapid growth and precocious maturation, whereas hooknoses grow to a larger body size and take longer to get there.

Thus, the problem of understanding the adult emergence of alternative male forms and behaviors has been pushed back to the problem of identifying the developmental factors that place newly hatched fish on one path or the other. Many possible factors could be involved in this process, including egg size, water flow, water temperature, and population density, as well as the identity of the parents.[82]

Similar to salmon, two separate male populations of blue-gill sunfish have been identified.[83] *Parental* male bluegills grow slowly and reach sexual maturity at seven years of age after attaining a large body size. At that time, they build a nest and guard it against intruder males. When a female enters the nest, she releases her eggs and the parental male releases his sperm to fertilize them. The survival of her fertilized eggs depends upon the protection provided by the parental male.

In contrast, *cuckolding* or *satellite* males reach sexual maturity at only two years of age and at a much smaller body size than parental males. Initially, they employ a sneaker strategy of fertilizing eggs—using stealth to dart into a nest, release sperm, and exit quickly. As these males continue to grow, they develop the coloration and behavior typical of female, not male bluegills. This female mimicry tricks parental males into letting down their guard, thus allowing satellite males to swim freely into nests and release sperm.

Whether employing a sneaker or satellite strategy, these cuckolds release their sperm and move on, leaving the parental male to guard the eggs, many of which have been fertilized by his competitors. For this reason, it is said that cuckolds parasitize parental males because they depend upon him to protect their offspring—the success of cuckolding depends upon the success of parenting.

Parental males seem to be playing the mating game by following the traditional male role, a perception that has led scientists to describe their behavior as *bourgeois*. They reproduce, as those old E. F. Hutton commercials used to say, the old-fashioned way—they *earn* it. They build a home, defend it, and protect their young from harm. The female and her eggs, by virtue of the male's hard work and investment of time and energy, *belong to him*.

But outside the parental male's home lurk larcenous males who seek all the benefits of offspring but none of the responsibilities. Scientists display their seeming disapproval for these males and their thieving ways when they refer to them as cuckolds, sneaks, and parasites. They have other names as well: scroungers, pseudofemales, transvestites.[84] I have even heard them referred to as sneaky fuckers. It seems that scientists have a soft spot for males that work hard and play by the rules.

But what are the rules? In bluegills, the cuckolds represent fully four-fifths of the breeding male population, a rather large proportion for an *alternative* mating strategy. What we have, really, are simply two mating lifestyles in evolutionary balance: Tip the scale toward more parental males and the benefits of a sneaking lifestyle increase, but tip the scale toward more sneakers and parental males—with the critical protection that they provide the eggs—are now favored. This balance arises evolutionarily as natural selection shapes the development of these fish within the context of their ecological circumstances.

As with the more familiar form of sexual differentiation that creates males and females, differentiation also occurs *within* the males of the species to produce two distinct life histories—beginning with differences in growth and the timing of sexual maturation, and ending with adult males that can seem as different from each other as each is different from females.[85] Indeed, as we have seen with bluegills, satellite males even mimic females in their pursuit of reproductive success.

The development of alternative males, comprising discernible forms and behaviors linked to distinct mating strategies, can seem like a gratuitous complication to an already complicated process. But perceiving such phenomena as complications arises from our preconceptions of how nature *ought* to behave—that sexuality has a dual essence and that all violations of that duality are, well, violations. However, as we have now seen, animals do not rest so easily on the conceptual bed that we have made for them.

There is a broader dynamic at work, one that captures sexual differentiation in all its forms. This dynamic entails not only the evolutionary construction of development, but the evolutionary construction of *alternative* developments[86]—alternatives that include the matching of an animal's form and behavior to the challenges that it is likely to face. Because the individual is calibrated developmentally, the match attained more accurately reflects current needs than if the individual's future were set in stone at the time of fertilization.

Sexual ambiguity may pose a problem for us, but it is no problem for bluegills, beetles, or bluehead wrasse, for anglerfish or tobaccofish, for salmon or Amazon mollies, or for hyenas or whiptail lizards. Each of these animals, and uncountable others, has elevated sexual oddity to evolutionary elegance. Once again, we see how our penchant for making absolute distinctions can clash with nature's tolerance for relative ones. But the true test of our acceptance of the lessons offered by these extraordinary animals is not whether we can tolerate their ambiguity, but whether we can embrace it.

MONSTROUS BEHAVIOR

We Still Have Much to Learn from
the Odd and Unusual

*If I want to see freaks,
I can just look out the window.*
JOHNNY ECK[1]

Franklin Dove's unicorn acquired, according to its creator, a "peculiar power." This bull used his single horn "as a prow to pass under fences and barriers in his path, or as a forward thrusting bayonet in his attacks."[2] Dove sensed the implications of the bull's seemingly novel behavior for our understanding of the causal roots of behavior: "The dominance of a unicorned animal over the ordinary two-horned beasts of the herd," Dove wrote, "is here offered as a striking instance of the dependence of behavior upon form."[3]

Dove's observation is anecdotal, but the point he raises—that bodily form shapes behavior—is valid. Johnny Eck, the two-legged goat, the oversized hind limbs of jerboas, and the star of the star-nosed mole all provide support for Dove's contention. Moreover, when the body is significantly altered and a behavioral change is induced, the brain is also altered significantly.

Like a peripheral device connected to a computer, the interface between appendage and brain must be compatible—and in the organic world, a successful interface is most easily and elegantly achieved when the whole system develops together. The result is a collection of integrated parts that seem made for each other.

One biologist has noted how "even a child can appreciate in a beast a sense of integrated wholeness, of design, of cooperation among its parts. In fact, these are the very attributes that somehow give animals their lovely and intriguing character."[4] We know who creationists look to—or rather, look up to—for the source of this design. For their part, evolutionists acknowledge only the *apparent* design of organisms and have countered the image of a celestial creator with the metaphor of a terrestrial tinkerer.[5]

I imagine this tinkerer—a child of the Modern Synthesis and its commitment to slow, incremental change—presenting his evolutionary handiwork as a slideshow, with aeons squeezed into each slide: *click*, the whale's four-legged ancestor re-enters the water; *click*, forelimbs transform into flippers; *click*, hindlimbs disappear. But what is missing from his presentation—what cannot be discerned at this great temporal distance—are the developmental details that made these evolutionary achievements possible. So as we come upon the sesquicentennial of Darwin's *Origin of Species*, it is time to insist on an evolutionary framework that is as exquisitely integrated as the animals that we seek to explain.

When Pere Alberch wrote of the logic of monsters,[6] he was alluding to internal developmental processes that shape, bias, and constrain the development of bodily form. For him, mon-sters—by which he meant anomalies of development—help us to see through the distracting glare of "design" to deeper truths *because monsters also exhibit an "integrated wholeness."*

But as long as monsters are banished to a province beyond the boundaries of traditional evolutionary thought, we are denied the benefits of their company. As Alberch, like William Bateson before him, understood, monsters provide a counterweight to the notion that natural selection is an omnipotent external force that shapes organisms without concern for the internal machin-ery of development. In a similar vein, Stephen Jay Gould and Richard Lewontin famously critiqued such *adaptationist* thinking thirty years ago. As they argued, our challenge is to appreci-ate "organisms as integrated wholes, fundamentally not decom-posable into independent and separately optimized parts."[7]

Alberch was concerned only with monstrous forms, not behaviors. But, as we have seen repeatedly throughout this book, the integrated wholeness of monstrous forms extends to their behavior. Johnny Eck, the two-legged goat, and armless wonders all testify to this integration—and jerboas, whales, snakes, and hyenas testify further to the integrative links among morphology, behavior, developmental anomalies, and evolutionary novelties. Thus, as a young jerboa's hindlimbs elongate and its body rises up to walk bipedally, we see not the emergence of some perfectly timed "bipedal instinct"—inbuilt and preprogrammed—but the unfolding of a developmental process through which brain and novel behavior are shaped.

Popular views of instincts mesh nicely with our sense that evolution has sculpted—and perfected—our brains to produce adaptive behavior.[8] But, truly, we cannot accurately assess the evolutionary influences on behavior if we restrict our focus only to well-formed animals living in ideal environments. Such a restricted focus only feeds illusions of ideal design. So, just as Alberch saw monsters as antidotes to adaptationist conceptions of form, monsters can also release us from the grip of similar illusions concerning behavior.

If *teratology* is the study of congenital malformations—a field founded by Etienne Geoffrey Saint-Hilaire and his son, Isidore—and *teratogeny* is the art of producing monstrous forms—a field first made feasible by Camille Dareste—then perhaps *terethology* is the appropriate designation for a new field: the study of the functional behavior of monsters. Why *terethology*? Just as ethology, the field founded in the middle of the last century by Konrad Lorenz and Niko Tinbergen, is devoted to studying adaptive behavior in its natural context, terethology would devote itself to studying behavior at the fringes of the natural world. To get the most out of this field, we would also want to investigate the brains of anomalous animals as well as their development. Thus, ultimately, the field of *developmental neuroterethology* would aim to provide a systematic framework for investigating the processes that integrate developing nervous systems with bodies packaged in novel forms to produce functional behaviors.

Such a science would not be completely new: After all, humans have been meddling with the workings of other animals (and of each other) for millenia—from selective breeding to genetic engineering—with the aim of refining behavior or exploring the mechanisms that produce it. But developmental terethology would be unique in its focus on the origins of functional behavior in strangely improbable yet fully organic creatures.

Of course I am aware of the fears that accompany any proposal that tolerates or promotes profound interference with nature. But where do these fears come from? How much have they been shaped by Drs. Moreau and Frankenstein and Hollywood horror films? As Stephen Jay Gould noted with more than a hint of sarcasm, horror movies consistently convey the message that our "cosmic arrogance can only lead to killer tomatoes, very large rabbits with sharp teeth, giant ants in the Los Angeles sewers, or even larger blobs that swallow entire cities as they grow."[9] Such fears reflect deep uncertainties about the consequences of human tinkering. But what kinds of creatures could humans possibly create that could be more astounding than those that nature has already produced?

In the New York Public Library—the same library where Callie Stephanides consulted a dictionary about her intersex condition—sits the 500-year-old copper Lenox Globe, particularly famous for its Latin phrase etched in a watery region off the eastern coast of Asia: *hic sunt dracones*. *Here be dragons.* Dragons and other monsters were commonly drawn in the distant and mysterious regions of maps and globes as warnings to sailors: Uncharted territory ahead. Perhaps today we can look upon our own vast globe and begin to embrace all that it has to offer—its wondrous diversity and its grotesque deformity—as we journey forth into an unknown future. Armless wonders and limbless snakes, elephants and cyclopic infants, blind mole-rats and star-nosed moles, "antlered" does and spotted hyenas, turtle shells and beetle horns, sneaking salmon and sex-switching tobaccofish, Abigail and Brittany Hensel, you and me and the person next door—here, everywhere, be monsters, not as antagonists but as protagonists together in one earthly drama.

ACKNOWLEDGMENTS

Numerous colleagues—some of whom I have never met—exhibited extraordinary generosity in responding to my unsolicited e-mails. I thank Pat Bateson, Jill Helms, Neil Shubin, Doug Emlen, Steve Glickman, Matthew Grober, David Gardiner, Frietson Galis, Scott Gilbert, Steven Black, Barry Mehler, Allen Enders, and Joseph-James Ahern of the American Philosophical Society. Scott Robinson donated many wonderful ideas. I am particularly grateful to Ann Burke for sharing with me her recollections of Pere Alberch, as well as Karen Adolph, David Lewkowicz, and five anonymous reviewers for taking time to read the manuscript and offer detailed, helpful comments.

Catharine Carlin of Oxford University Press took me to lunch several years ago and we have been working together ever since. Her confidence in this book, and her friendship, made all the difference. Erin Nelson assisted me with inspiring energy and creativity in the early stages of this project and was instrumental in helping me get this book off the ground. My agent, Ethan Ellenberg, again guided me wisely. I also appreciate the support of Latha Menon of Oxford University Press. Ronnie Lipton and Marion Osmun are wonderful editors and I cannot overstate my gratitude to them for their guidance, skill, and insight.

Family, friends, colleagues, and students provided much-needed support through good times and bad. I thank them all. Finally, once again, I thank my wife Jo McCarty for graciously enduring the seemingly endless writing process—reading, listening, and advising with the patience of a saint, the insight of a psychoanalyst, and the encouragement of a coach. Words alone do not suffice…

NOTES

Introduction

1. See Jacob (1977), p. 1163. Interestingly, Jacob begins his article on *Evolution and Tinkering* with a discussion of monsters and how they have been rendered by different cultures. By examining such renderings, he writes, we can see how "a culture handles the possible and marks it limits" (p. 1161).

2. In Jacob's (1977) original discussion of the tinkerer, he states that evolution "behaves like a tinkerer who, during eons upon eons, would slowly modify his work, unceasingly retouching it, cutting here, lengthening there, seizing the opportunities to adapt it progressively to its new use" (p. 1164). Elsewhere, he writes that natural selection works on "random variations" and thereby "gives direction to changes, orients chance, and slowly, progressively produces more complex structures, new organs, and new species" (p. 1163).

3. See Provine (1971).

4. Bateson (1894/1992), p. 19.

5. Lewontin (1992, pp. 137–138) writes, "The essential feature of Mendelism is a causal rupture between the processes of inheritance and the processes of development. What is inherited, according to Mendelism, is the set of internal factors, the genes, and the internal genetic state of any organism is a consequence of the dynamical laws of those entities as they pass from parent to offspring. Those laws of hereditary passage contain no reference to the appearance of the organism and do not depend upon it in any way."

6. Kirschner and Gerhart (2005), p. 31.

7. See Gilbert's (2003) review of the historical roots of contemporary evolutionary developmental biology (or Evo Devo). See also Alberch (1989) and Gottlieb (1992).

8. For a recent exploration of the mechanisms that produce variability, see Kirschner and Gerhart (2005). For an account of William Bateson's place within Evo Devo, see Carroll (2005).

9. West-Eberhard (2003), p. 193.

10. Ibid., p. 208.

11. Ibid., p. 205.

12. See Kirschner and Gerhart (2005), p. 37.

13. Rachootin and Thomson (1981), p. 181.

14. For a fuller discussion of "designer thinking," see Chapter 2 in Blumberg (2005).

1. A Parliament of Monsters

1. Bateson (1908), p. 19.

2. The original reports describing the anatomy and biomechanics of the two-legged goat were from Slijper (1942a,b). For subsequent discussions of the evolutionary significance of this phenomenon and

others like it, see Rachootin and Thomson (1981), Alberch (1982), and West-Eberhard (2003).

3. See O'Connor (2000), p. 148.

4. Arbus, *Revelations* (2003), p. 57.

5. Ibid., p. 67.

6. Stevenson (1993).

7. Ibid.

8. Bates (2005), p. 158.

9. Bates (2000).

10. See Bates (2005), p. 145ff.

11. Ibid., pp. 3–4.

12. Paré (1573/1982), p. 3.

13. See Bates (2005), p. 13.

14. Ibid., p. 67.

15. Ibid., p. 16.

16. Ibid., p. 11.

17. Ibid., p. 80.

18. Ibid., p. 81.

19. Quoted in Oppenheimer (1968), p. 147.

20. Gould and Pyle, (1896), p. 193.

21. Ibid., p. 301.

22. Ibid., p. 1.

23. However, as noted by McNamara (1997, p. 30), "Darwin seems to have been remarkably equivocal about the whole question of the importance of embryology to evolution. While, on the one hand, he wrote in *The Origin of Species* (6th edition) that 'embryology will often reveal to us the structure, in some degree obscured, of the prototypes of each great class,' he devoted a mere 10 out of 429 pages to the role of embryology in evolution. Perhaps Darwin was merely reacting against

what had been the all-pervasive influence of embryological studies to species interrelationships in the decades leading up to the publication of *The Origin of Species.* Yet he clearly recognized the importance of embryonic development to classification, noting that 'community in embryonic structure reveals community of descent.'"

24. Huxley (1863), p. 57.

25. When Darwin, in *The Origin of Species* (p. 599), concluded that "community in embryonic structure reveals community of descent," he was suggesting that embryonic development can be used to track the path of evolution, reflecting his apparent acceptance of Ernst Haeckel's infamous notion that "ontogeny recapitulates phylogeny." Like Haeckel, Karl Ernst von Baer was a nineteenth century embryologist who also closely examined parallels between ontogeny and phylogeny. However, writing before Haeckel, von Baer formulated a developmental framework that excluded recapitulation as a viable notion and embraced the epigenetic perspective of development as a transformational process. It was only later, in the twentieth century, that embryologists would finally put Haeckel's ideas to rest. As the embryologist Walter Garstang (1922, p. 82) forcefully argued, "Ontogeny does not recapitulate phylogeny: it creates it." For reviews and related discussion, see Gould (1977), Gottlieb (1992), McNamara (1997), Richardson et al. (1997), and Richardson (1999).

26. Huxley (1863), p. 65.

27. For example, as Gottlieb (1992) relates, Garstang opened "the door for the notion that all sorts of changes in ontogeny (not only additions at the end of development) were the basis of evolutionary change....In essence, what Garstang did was to put von Baer's non-evolutionary generalizations concerning ontogeny into an evolutionary framework...." (pp. 90–91). The embryologist Gavin de Beer also rejected terminal addition as the sole mechanism of evolutionary change,

writing in *Embryos and Ancestors* (1940) that "evolutionary novelties can and do appear in *early* stages of ontogeny" (p. 32).

28. Garstang (1922), p. 84, quoted in Gottlieb (1992), p. 93.

29. Richardson et al. (1997).

30. Richardson (1999).

31. See Richards (1992) and Gottlieb (1992).

32. Darwin's interest in domesticated animals and the "art" of breeding achieved its fullest expression in *The Variation of Animals and Plants Under Domestication,* published in 1868.

33. Darwin (1859/1983), p. 595.

34. Bateson (1894/1992).

35. According to Gottlieb (1992) St. George Mivart "appears to be the first systematic thinker to make individual ontogenetic development central to his view of the basis for evolutionary change" (p. 40). Writing several decades before Bateson, Mivart also discounted natural selection as the sole evolutionary force and disputed the Darwinian commitment to continuity. As we examine evolutionary thinkers over the past two centuries, it appears that a focus on internal, developmental factors encourages acceptance of the possibility of discontinuous change.

36. Ibid., p. 6.

37. Ibid., p. 17.

38. Huxley believed that Darwin's commitment to continuous variation was "an unnecessary difficulty" that he had placed on himself; see Provine (1971), p. 12.

39. See Richards (1994), pp. 407–411.

40. In *The Variation of Animals and Plants under Domestication,* Darwin repeatedly concerns himself with monstrosities, their origins, and their implications for his theory of evolution. For example, after noting that some monstrous breeds of dogs "have probably arisen suddenly" and

may be "fixed by man's selection," he quickly adds "that the most potent cause of change has probably been the selection, both methodical and unconscious, of slight individual variations—the latter kind of selection resulting from the occasional preservation, during hundreds of generations, of those individual dogs which were the most useful to man for certain purposes and under certain conditions of life" (p. 40).

41. Contrary to the joke that Goldschmidt has become in some evolutionary quarters, he was a serious thinker and respected geneticist whose writings reflect a keen appreciation for the central role played by development in evolutionary change (see Bush, 1982, for one biologist's recollection of the ridicule heaped on Goldschmidt for his unorthodox ideas). His *hopeful monsters* do not simply pop into existence—they develop. In his words: A "genetic change involving rates of embryonic processes does not affect a single process alone. The physiological balanced system of development is such that in many cases a single upset leads automatically to a whole series of consecutive changes of development in which the ability for embryonic regulation, as well as purely mechanical and topographical moments, come into play; there is in addition the shift in proper timing of integrating processes. If the result is not, as it frequently is, a monstrosity incapable of completing development or surviving, a completely new anatomical construction may emerge in one step from such a change" (1940, p. 386). "Species differences," he wrote, "are differences of the whole developmental pattern" (p. 181).

In his introduction to Goldschmidt's *The Material Basis of Evolution*, Gould (1982) provides a balanced presentation of Goldschmidt's "heretical" ideas, including his wrong turns. Frazzetta (1970) invokes Goldschmidt to help explain the seemingly discontinuous evolution of the distinct jaw of Bolyerine snakes. Addressing the question of how a monster could compete successfully in the struggle for existence, Frazzetta writes, "An individual with a greatly revised morphology, suddenly appearing in a population, may seem to have next to no chance of surviving alongside of others whose adaptive systems have been gradually tuned and integrated by generations of selection. But it need not follow that a monstrous individual lacks adaptive integration of the modified system. Organic structures display a remarkable capacity for functional adjustment" (p. 66).

42. The notion of the *hopeful monster* continues to evoke powerful reactions (regardless of how faithfully that concept is lifted from Goldschmidt's original writings). For example, after Olivia Hudson wrote a newspaper article entitled "The monster is back, and it's hopeful" (http://judson.blogs.nytimes.com/2008/01/22/the-monster-is-back-and-its-hopeful/), Jerry Coyne quickly issued a sharp rejoinder (http://scienceblogs.com/loom/2008/01/24/hopeless_monstersa_guest_post.php). It is difficult to judge such disputes, especially when there are so many complex threads coursing through each argument. For example, as this book attests, I share Hudson's fascination with monsters and their implications for evolution, but I am not comfortable with her or Coyne's exclusive focus on genes and mutations—as opposed to epigenetic systems that can change as a result of mutation or environmental factors. And, Coyne's unwavering commitment to gradualism (in the traditional Darwinian sense) makes me uneasy. Regardless, by the end of such disputes, including that between Bateson and the Darwinians, it becomes clear that continuity and discontinuity are relative terms that are best set aside so that we can focus on actual empirical findings.

43. For a detailed history of William Bateson's dispute with the biometricians and the rise of the Modern Synthesis, see Provine (1971).

44. Quoted in Provine (1971), p. 51.

45. Lewontin (1992) attributes this "rupture" between inheritance and development to the rise of Mendelism. West-Eberhard (2003) emphasizes that Darwin "had already recognized this distinction and

was struggling to explain both" (p. 192). Indeed, as she argues, "Darwin avoided the false dichotomy between nature and nurture by insisting that inheritance includes both the transmission and expression of traits" (p. 188). Thus, it "has taken more than 120 years to come back to where Darwin left off...." (p. 193).

46. See Gilbert (2003). See also Gottlieb (1992) for a discussion of the founders of the Modern Synthesis and their failure "to incorporate individual development into evolutionary theory." He writes, "The brilliant architects of the modern synthesis were certainly not ignorant of the role of gene action in development, it is just that population-genetic thinking made them prone to put the analysis of evolution in terms of heritability, genes, and environment rather than in developmental terms, something that remains with us today, as population-genetic thinking is still the dominant mode of contemporary biological thought concerning evolution" (p. 126).

47. Kirschner and Gerhart (2005), p. 29.

48. For a review of the historical roots of Evo Devo, see Gilbert (2003).

49. Carroll (2005), p. 4.

50. For further discussion, see Keller (2000), Moore (2001), and Blumberg (2005).

51. Leroi (2003), p. xiii.

52. See Nijhout (1990).

53. Interview with Ann Burke, March 6, 2006.

54. Alberch (1989), p. 27.

55. It should be noted that Alberch does not cite William Bateson in this paper. However, he does cite Richard Goldschmidt as one who appreciated the importance of internal factors in evolution.

56. Alberch (1989) describes his "internalist scheme" as follows: "Perturbations resulting from genetic mutation or environmental impact are 'filtered' through the dynamics of a pattern-generating system. The

internal structure of the developmental system defines a finite and discrete set of possible outcomes (phenotypes) even if the sources of perturbation are random" (p. 27). His interest in extreme variants, that is, monsters or teratologies, was "as model systems to study the pattern generated by developmental properties. Unlike the classical 'hopeful monsters' of Goldschmidt (1940), I do not contend that teratologies are variation [sic] with evolutionary potential. In fact, I assume them to be lethal in most cases" (p. 28).

57. Alberch (1989), p. 48.

58. The behavioral embryologist, Gilbert Gottlieb, refers to selection acting on the entire *developmental manifold* (see Gottlieb, 1997). For an historical review of *epigenesis*, its opposite, *preformationism*, and the schools of thought that flowed from these ideas, see Gottlieb (1992).

59. See Gottlieb (1997).

60. See Jablonka and Lamb (2002).

61. Today, you may read that molecular biologists have founded a "new" science of epigenetics. In that context, *epigenetics* refers to the regulation of DNA by extragenetic mechanisms (including what is unfortunately referred to as the "epigenetic program"). These extragenetic mechanisms fall within the rubric of the classical notion of *epigenesis*. However, *epigenesis* is broader than any single mechanism and encompasses all of the processes that shape development.

62. Waddington (1953).

63. See Oyama et al. (2001).

64. West-Eberhard (2005), p. 6547.

65. Wells (1896/2004), p. 68.

66. Ibid.

67. Ibid., p. 57.

68. There are many examples of such bizarre animals created through embryonic tinkering. For example, Constantine-Paton and Law (1978)

created frogs with a third eye situated between the other two and used this creature to examine the principles guiding the neural development of the visual system. More recently, cells from embryonic ducks have been transplated into embryonic quail ('qucks'), and vice versa ('duails'), to produce hybrids that make possible the investigation of those epigenetic factors that guide the development of species-typical beaks and feathers. As Schneider (2005) notes, the inductive changes instigated by the donor cells upon transplantation into the embryonic host are "accepted with acquiescent agility" (p. 570).

69. See Graham et al. (1993).

2. *Arresting Features*

1. Stockard (1921), p. 117.

2. de Beer (1940), p. 21.

3. http://news.nationalgeographic.com/news/2003/02/0205_030205_cyclops.html.

4. See Mayor (2000), for an examination of the link between fossils and Greek and Roman mythology.

5. See Roessler and Muenke (1998).

6. Barr and Cohen (1999).

7. Wilder (1908), p. 356.

8. Ibid., p. 358.

9. Ibid., p. 359.

10. Li et al. (1997).

11. Gaeth et al. (1999).

12. Cooper (1998).

13. Bendersky (2000).

14. Lesbre (1927), p. 248.

15. Wilder (1908), p. 356.

16. Ibid., p. 367.

17. Ibid.

18. Ibid., p. 368.

19. Ibid., p. 427.

20. Ibid.

21. Ibid., p. 436.

22. Rats are insensitive to red light, thus allowing humans to observe their behavior in the dark.

23. See Wilson (2003), p. 21.

24. For example, one of Stockard's articles was entitled "An experimental study of racial degeneration in mammals treated with alcohol."

25. Pearl (1916), p. 255.

26. Pauly (1996), pp. 12–13.

27. Quoted in Pauly (1996), p. 15.

28. See Pauly (1996) for an account of this history.

29. Quoted in Pauly (1996), pp. 19–20.

30. Pearl (1927), p. 260.

31. Comments of Haven Emerson, Conference on Medicine and Eugenics held at the New York Academy of Sciences, April 21, 1937.

32. Comments of Charles Stockard, Conference on Medicine and Eugenics held at the New York Academy of Sciences, April 21, 1937.

33. See Stockard (1941).

34. See Oppenheimer (1968).

35. Quoted in Oppenheimer (1968), p. 150.

36. Ibid., pp. 145–146.

37. Gerhart et al. (1986), p. 541.

38. Stockard (1921), p. 139.

39. Ibid., p. 120.

40. Ibid., p. 152.

41. Ibid., p. 245.

42. Ibid., p. 247.

43. Roessler and Muenke (1998).

44. For a detailed discussion of interchangeability, see Chapter 5 in West-Eberhard (2003). Richard Goldschmidt coined the term *phenocopy* to describe a creature that "copies the type of a mutant" (1940, p. 267) but is produced through an environmentally induced alteration of development. It is the interchangeability of genetic and environmental mechanisms that makes phenocopies possible.

45. Barr and Cohen (1999).

46. Binns et al. (1963).

47. Keeler and Binns (1968); Keeler (1970).

48. Chiang et al. (1996); Cooper et al. (1998).

49. I use the convention here of referring to a gene in italics (e.g., *Shh*) and the protein associated with that gene in non-italics (e.g., Shh).

50. For example, Carroll (2005).

51. Cordero et al. (2004), p. 485.

52. Ibid.

53. In *Embryos and Ancestors* (1940), Gavin de Beer devotes his third chapter to "The speeds of the processes of development." He ends this chapter in this way: "Altogether, then, we may safely conclude that the speeds at which the internal factors [i.e., genes] work are of great importance in development, and variations in the relative speeds of the various factors may play an important part in the relation of ontogeny to phylogeny" (p. 21). Similarly, as already noted, Richard Goldschmidt, in *The Material Basis of Evolution* (1940), placed great emphasis on genes as regulators of the timing of development.

54. Ibid., p. 487.

55. Carles et al. (1995).

56. Phelan (1993); Kaufman (2004).

57. See Kaufman (2004).

58. See Spencer (2000, 2003).

59. http://abcnews.go.com/Primetime/story?id=2346476&page=1.

60. Wewerka and Miller (1996).

61. Black and Gerhart (1986).

62. Ulshafer and Clavert (1979).

63. Stockard (1921), p. 188.

64. Ibid., p. 251.

65. West et al. (1988).

66. Kirschner and Gerhart (2005), pp. 34–35.

67. West-Eberhard (2005), p. 6547; see also Jablonka and Lamb (2005).

68. West-Eberhard (2003), p. 20.

69. Gottlieb (1992), p. 177.

70. Ibid., p. 176.

3. Do the Locomotion

1. He actually possessed two undeveloped legs that he kept hidden inside his clothes, a condition known as *amelia*.

2. Humphrey et al. (2005).

3. Karen Adolph, personal communication.

4. For a description of the "bear crawl," see Adolph et al. (1998) and Burton (1999). Adolph et al. (in press) discuss this form of locomotion in relation to the Turkish hand-walkers.

5. Faith has a web site: http://www.faiththedog.net/index.asp

6. See West-Eberhard (2003), pp. 51–52.

7. Rachootin and Thomson (1981), p. 184.

8. West-Eberhard (2005), p. 6545; for a similar perspective on the evolution of bipedal walking, see Hirasaki et al. (2004).

9. See Jablonka and Lamb (2005).

10. For example, Alberch (1989, p. 28) writes: "Unlike the classical 'hopeful monsters' of Goldschmidt (1940), I do not contend that teratologies are variation [sic] with evolutionary potential. In fact,

I assume them to be lethal in most cases." Similarly, West-Eberhard (2003, p. 52) states that the point of examples like the two-legged goat "is not to argue that these handicapped individuals might have given rise to durable novelties, for it is unlikely in the extreme that such individuals would outperform normal individuals in nature."

11. Alberch (1982), p. 27.

12. West-Eberhard (2003, p. 51) defines *phenotypic accommodation* as "adaptive mutual adjustment among variable parts during development without genetic change."

13. Rachootin and Thomson (1981), p. 186.

14. See Alexander (1992) for a general introduction to biomechanics.

15. Bramble and Lieberman (2004).

16. See Hoyt and Taylor (1981).

17. See Adolph et al. (1998) and Adolph and Berger (2006).

18. See Adolph and Joh (2007).

19. Blumberg-Feldman and Eilam (1995); Eilam (1997); Eilam and Shefer (1997).

20. Viala et al. (1986).

21. For related descriptions of motor learning in fetal rats, see Robinson (2005) and Robinson and Kleven (2005).

22. For a recent discussion of core knowledge, see Spelke and Kinzler (2007). For a spirited critique of that concept, see Spencer et al. (in press).

4. Life and Limb

1. Goldschmidt (1933), p. 544.

2. Quoted in O'Connor (2000), p. 198.

3. Ibid., p. 102ff.

4. Carmena et al. (2003); Lebedev and Nicolelis (2006).

5. See Flor et al. (1998) and Lotze et al. (1999) for discussions

of phantom limb, phantom limb pain, and cortical reorganization after amputation.

6. O'Connor (2000), p. 198.

7. See Bates (2005), p. 218.

8. Flor et al. (1998).

9. Documentary entitled *Smiling in a Jar*, featuring an interview with Gretchen Worden, Director of the Mütter Museum. The documentary was originally televised as part of a series entitled *Errol Morris' First Person* (2000).

10. Brenner (1992).

11. For example, see Krubitzer and Kaas (2005).

12. For example, Roder et al. (1999).

13. Cohen et al. (1997).

14. Kahn and Krubitzer (2002).

15. To produce these maps, animals are anesthetized and the surface of their cerebral cortex is exposed. As a recording electrode is moved systematically across the surface, the investigator stimulates the animal in a variety of ways: touching the skin surface throughout the body, shining light into the eyes, making sounds, etc. When all the information is tallied, a map relating each piece of cortical tissue to a sensory system is produced. Next, it has become popular to convert these maps into cartoon representations of the animals to illustrate the relative significance of each sensory system. Such cartoons are presented here in the figures for the naked mole-rat and the star-nosed mole.

16. Bronchti et al. (2002).

17. Heil et al. (1991).

18. For review, see Catania and Henry (2006).

19. Catania and Kaas (1997).

20. Catania and Remple (2005).

21. Catania and Kaas (1997).

22. See Killackey et al. (1995).

23. Ibid.

24. Elbert et al. (1995); see also Clark et al. (1998).

25. Krubitzer (1995), p. 415.

26. Cortical reorganization after limb amputation can even occur in adults, as has been demonstrated in human (Elbert et al., 1997) and nonhuman (Merzenich et al., 1984) animals.

27. See Burke and Nowicki (2001).

28. See Cohn and Tickle (1999).

29. For review, see Tickle (2002).

30. Alberch and Gale (1985).

31. Ibid.

32. See Figure 4 in Oster et al. (1988).

33. See Alberch and Gale (1985).

34. Alberch (1985).

35. Alberch and Gale (1985), p. 19.

36. See Newman and Müller (2000) for a recent, extended discussion of epigenesis and evolutionary innovations.

37. Cohn and Tickle (1999).

38. Bejder and Hall (2002).

39. Thewissen et al. (2006).

40. Ibid., Cohn and Tickle (1999).

41. Bryant et al. (2002); Galis et al. (2003).

42. Bryant et al. (2002), p. 895.

43. McGann et al. (2001).

44. See Galis et al. (2003).

45. Ibid.

46. See Blaustein and Johnson (2003).

47. Ibid., p. 88.

48. Ibid., Johnson et al. (1999); Kiesecker (2002).

49. Johnson et al. (1999); Stopper et al. (2002); Johnson et al. (2006).

50. Stopper et al. (2002).

51. Johnson et al. (2006).

52. Burke (1989).

53. Cebra-Thomas et al. (2005).

54. Gilbert et al. (2001), p. 56.

55. Carroll (1960), pp. 286–287.

56. Dove (1936), p. 435.

57. Emlen (2000), p. 407.

58. See Moczek et al. (2006).

59. Hölldobler and Wilson (1990).

60. Hunt and Simmons (2000).

61. Ibid.

62. See Jablonka and Lamb (2005) for an extended analysis of heredity in all its forms.

63. See Emlen (2000).

64. Ibid., p. 412.

65. Ibid.

66. Quoted in Le Guyader (1998), p. 22.

67. Stockard (1921), p. 118.

68. Emlen (2001).

69. Emlen (2000), p. 415.

70. Alberch (1989), p. 41.

5. Anything Goes

1. Eugenides (2002), p. 520.

2. It is Ann Burke who first identified the carapacial ridge in turtles and suggested that it played a role in the evolution of the shell, as discussed in Chapter 4.

3. Interview with Ann Burke, March 6, 2006.

4. See Dreger (1998).

5. Eugenides (2002), p. 431.

6. Ibid., p. 433.

7. Ibid., p. 437.

8. Ibid., p. 479.

9. Ibid.

10. Quoted in Fausto-Sterling (2000), p. 47.

11. Quoted in Rosenberg (2007), p. 56.

12. For example, see Fischer (1970).

13. Gilbert and Zevit (2001).

14. Male circumcision is practiced in many cultures on infants, adolescents, and adults. It is justified by religious practice and supposed hygiene benefits and leaves the penis in a functioning state. By contrast, female "circumcision" entails complete removal of the clitoris and, very often, the surrounding tissue. This particularly venal form of genital mutilation is perpetrated on millions of young girls every year, primarily but not exclusively in Africa, resulting in life-long physical and psychological trauma as well as death. The United Nations Convention on the Rights of the Child has condemned this practice. See Black (1997).

15. For review, see Dreger (1998).

16. Diamond and Sigmundson (1997).

17. See Colapinto (2004).

18. McHugh (2004).

19. McHugh (1995), p. 111.

20. Ibid.

21. Ibid.

22. McHugh (2004).

23. Ibid.

24. For example, see the recent *Newsweek* cover story entitled *Rethinking gender* (Rosenberg, 2007).

25. See Fausto-Sterling (2000), p. 53.

26. See Houk et al. (2006).

27. Dreger (1998).

28. See Fausto-Sterling (2000), pp. 56–63.

29. Woodhouse (1998), p. 9.

30. Reiner and Gearhart (2004).

31. McHugh (2004).

32. For example, see Fausto-Sterling (2000), p. 79ff, and Weil (2006).

33. Bin-Abbas et al. (1999).

34. Fagot and Leinbach (1989).

35. See Bem (2000).

36. Ibid., p. 533.

37. For a recent example of a biological explanation for male sexual orientation that does not depend upon genetic differences, see Bogaert (2006). Specifically, it was found that males with more older brothers were more likely to be homosexual, perhaps due to progressive changes in the gestational environment provided by the mother. Based on these and other findings, some now believe that the developmental path to homosexuality (at least in males) begins during the prenatal period. For a brief introduction to these ideas and the related literature, see Puts et al. (2006).

38. See Blumberg (2005) for a review of the instinct concept.

39. Mayr (1976).

40. See ten Cate (1994) for review.

41. See West et al. (1988); Oyama et al. (2001); Jablonka and Lamb (2005); Blumberg (2005).

42. De Guise et al. (1994).

43. Scanlon et al. (1975).

44. Fahrenthold (2006).

45. Wiig et al. (1998); Sonne et al. (2005) have suggested that the enlarged clitorises of these polar bears could have been due to inflammation caused by excessive licking, thus excluding them as intersexes. Nonetheless, scientists continue to find evidence in the Arctic of increased levels of environmental contaminants with hormone-like effects that could explain the increased incidence of apparent intersex conditions in polar bears (for example, see Muir et al., 2006).

46. Stamper et al. (1999).

47. For references, see Cattet (1988).

48. The male's Wolffian system includes the vas deferens, seminal vesicles, and epididymis. The female's Mullerian system includes the uterus and fallopian tubes.

49. Hunter (1779), p. 285.

50. For a review of Lillie's contributions, see Capel and Coveney (2004).

51. Lillie (1916), p. 612.

52. Shozu et al. (1991); Conte et al. (1994).

53. Typically, the initiation of sex differentiation can be traced to the sex-determining region on the Y chromosome (SRY) that codes for the SRY protein. This protein, when present before the second month of gestation in human fetuses, induces the transition of the ovotestis to a testis. The SRY protein itself functions like a hormone, so it can be administered like one: It can be injected into XX fetuses to stimulate the formation of testes. Or agents can be injected into XY fetuses that block the SRY protein's action and thereby prevent the formation of testes there by. Or the SRY gene can be transferred to any chromosome in an XX fetus, resulting in the production of the SRY protein and, ultimately, the creation of testes. Such translocations of the SRY gene occur in the majority (90%) of human cases of so-called "XX sex reversal,"

an intersex condition characterized by the presence of testes despite the absence of a Y chromosome. In addition, the formation of testes in XX individuals that also lack an *SRY* gene is typical in some species (e.g., dogs); this phenomenon is still poorly understood. See Vaughan et al. (2001) and Kuiper et al. (2005).

54. For review, see Modi and Crews (2005).

55. See Crews et al. (1995).

56. See Modi and Crews (2005).

57. Quinn et al. (2007).

58. Sarre et al. (2004), p. 641.

59. For review, see Vandenbergh (2003).

60. Cattet (1988).

61. Ibid., p. 849.

62. For a review of the hyena's reputation through history, see Glickman (1995).

63. For review, see Glickman et al. (2006).

64. Although administering androgen-blocking agents to female fetuses does not prevent the formation of an enlarged clitoris, it does make the female form even more feminine (for example, shorter and thicker). This finding suggests that female fetuses are exposed to some androgens *in utero* that slightly masculinize their already enlarged clitorises. See Glickman et al. (2006) for an extended discussion of this issue.

65. See Place and Glickman (2004), for some preliminary clues. Glickman (personal communication, June 18, 2007) believes that estrogens (likely produced by the maternal ovary) are responsible for the early fetal enlargement of the hyena phallus.

66. Munday et al. (2005); Black and Grober (2003).

67. Petersen (1995).

68. This principle was first described by Michael Ghiselin and is

known as the size advantage model for sequential hermaphroditism. See Godwin et al. (2003), p. 41. In contrast, Rodgers et al. (2007) argue that social status, rather than size, guides sexual transformations in fish.

69. Munday et al. (2005).

70. Munday et al. (2006).

71. Semsar and Godwin (2003).

72. Munday et al. (2006), p. 2845.

73. For review, see Pietsch (2005).

74. Amazon mollies and their related species fertilize their eggs internally and give birth to live young. In other species (e.g., tobaccofish), the eggs are fertilized externally.

75. See Neiman (2004) for a discussion of unisexual species and their dependence on the copulatory mechanisms that characterize their sexual ancestral species.

76. For a review of the regulatory control of sexual behavior in whiptail lizards, see Crews (2005).

77. Moczek and Emlen (2000).

78. See Gross (1996).

79. For review, see Gross (1991).

80. Technically, the terms *jack* and *hooknose* are specific to coho salmon. Atlantic salmon exhibit similar life histories as coho salmon, but scientists refer to the small and large males as *precocious parr* and *adults*, respectively.

81. Metcalfe et al. (1989); Thorpe et al. (1992); Aubin-Horth and Dodson (2004).

82. For a recent look at this question in Atlantic salmon, see Aubin-Horth and Dodson (2004).

83. Gross and Charnov (1980); Neff et al. (2003).

84. See Taborsky (1994) for a thorough accounting of the many synonyms for sneakers and their ilk.

85. Moore (1991); Gross (1996).

86. See Caro and Bateson (1986) for a detailed discussion of the need for developmental analyses for understanding alternative reproductive strategies.

Epilogue: Monstrous Behavior

1. The story behind this comment, accessed from http://phreeque. tripod.com/johnny_eck.html, is as follows: "The event that turned Johnny from a beloved local celebrity into a sullen old recluse was a robbery at the family home, which he and [his brother] Robert inhabited, in 1987. Old and enfeebled, Johnny was unable to defend himself as a gang of thieves physically restrained him and walked off with his valuables. It was this incident that is said to have inspired his famous quote, 'If I want to see freaks, I can just look out the window,' indicating that the once-congenial King of Freaks had finally lost faith in his fellow man. On January 5, 1991, after almost four years of living in total seclusion, Johnny suffered a heart attack and died."

2. Dove (1936), p. 435.

3. Ibid., p. 436.

4. Frazzetta (1975), p. 2.

5. Jacob (1977).

6. Alberch (1989).

7. Gould and Lewontin (1979), p. 591.

8. See Blumberg (2005) for an extended discussion of the instinct concept and its pitfalls.

9. Gould (1995), p. 53.

SOURCES AND SUGGESTED READING

Adelmann, H. B. 1936. The problem of cyclopia. Part I. *Quarterly Review of Biology* 11, 161–182.

———. 1936. The problem of cyclopia. Part II. *Quarterly Review of Biology* 11, 284–304.

Adolph, K. E., & Berger, S. E. 2006. Motor development. In *Handbook of child psychology*: Volume 2: *Cognition, perception, and language* (pp. 116–213), D. Kuhn & R. S. Siegler (Eds.) New York: Wiley.

Adolph, K. E., & Joh, A. S. 2007. Motor development: How infants get into the act. In *Introduction to infant development* (pp. 63–80), A. Slater & M. Lewis (Eds.) New York: Oxford University Press.

Adolph, K. E., Karasik, L. B., & Tamis-LeMonda, C. S. (in press). Moving between cultures: Cross-cultural research on motor development. In *The handbook of cross-cultural developmental science*, M. H. Bornstein (Ed.) Mahwah, NJ: Lawrence Erlbaum.

Adolph, K. E., Vereijken, B., & Denny, M. A. 1998. Learning to crawl. *Child Development* 69, 1299–1312.

Alberch, P. 1980. Ontogenesis and morphological diversification. *American Zoologist* 20, 653–667.

———. 1982. The generative and regulatory roles of development in evolution. In *Environmental adaptation and evolution* (pp. 19–36), D. Mossakowski & G. Roth (Eds.) Stuttgart: Gustav Fischer.

———. 1983. Mapping genes to phenotypes, or the rules that generate form. *Evolution* 37, 861–863.

———. 1985. Developmental constraints: Why St. Bernards often have an extra digit and poodles never do. *American Naturalist* 126, 430–433.

———. 1989. The logic of monsters: Evidence for internal constraint in development and evolution. *Geobios* 19, 21–57.

Alberch, P., & Gale, E. A. 1985. A developmental analysis of an evolutionary trend: Digital reduction in amphibians. *Evolution* 39, 8–23.

Alberch, P., Gould, S. J., Oster, G. F., & D. B. Wake. 1979. Size and shape in ontogeny and phylogeny. *Paleobiology* 5, 296–317.

Alexander, R. M. 1992. *Exploring biomechanics*. New York, Scientific American Library.

Andreadis, P. T., & Burghardt, G. M. 1993. Feeding behavior and an oropharyngeal component of satiety in a two-headed snake. *Physiology & Behavior* 54, 649–658.

Arbus, D. 2003. *Revelations*. New York: Random House.

Aubin-Horth, N., & Dodson, J. J. 2004. Influence of individual body size and variable thresholds on the incidence of a sneaker male reproductive tactic in Atlantic salmon. *Evolution* 58, 136–144.

Barr, M., Jr., & Cohen, M. M., Jr. 1999. Holoprosencephaly survival and performance. *American Journal of Medical Genetics* 89, 116–120.

Bates, A. W. 2000. Birth defects described in Elizabethan ballads. *Journal of the Royal Society of Medicine* 93, 202–207.

———. 2005. *Emblematic monsters: Unnatural conceptions and deformed births in early modern Europe.* Amsterdam: Rodopi.

Bateson, P. 2002. William Bateson: a biologist ahead of his time. *Journal of Genetics* 81, 49–58.

Bateson, W. 1894/1992. *Materials for the study of variation.* Baltimore: Johns Hopkins University Press.

———. 1908. *The methods and scope of genetics.* Cambridge: Cambridge University Press.

Bejder, L., & Hall, B. K. 2002. Limbs in whales and limblessness in other vertebrates: mechanisms of evolutionary and developmental transformation and loss. *Evolution & Development* 4, 445–458.

Bem, D. J. 2000. Exotic becomes erotic: Interpreting the biological correlates of sexual orientation. *Archives of Sexual Behavior* 29, 531–548.

Bendersky, G. 2000. Tlatilco sculptures, diprosopus, and the emergence of medical illustrations. *Perspectives in Biology and Medicine* 43, 477–501.

Bin-Abbas, B., Conte, F. A., Grumbach, M. M., & S. L. Kaplan. 1999. Congenital hypogonadotropic hypogonadism and micropenis: effect of testosterone treatment on adult penile size why sex reversal is not indicated. *Journal of Pediatrics* 134, 579–583.

Binns, W., James, L. F., Shupe, J. L., & G. Everett. 1963. A congenital cyclopian-type malformation in lambs induced by maternal ingestion of a range plant, *Veratrum californicum. American Journal of Veterinary Research* 24, 1164–1175.

Black, J. 1997. Female genital mutilation: a contemporary issue, and a Victorian obsession. *Journal of the Royal Society of Medicine* 90, 402–405.

Black, M. P., & M. S. Grober. (2003). Group sex, sex change, and parasitic males: Sexual strategies among the fishes and their neurobiological correlates. *Annual Review of Sex Research.* 14, 160–184

Black, S. D., & Gerhart, J. C. 1986. High-frequency twinning of *Xenopus laevis* embryos from eggs centrifuged before first cleavage. *Developmental Biology* 116, 228–240.

Blaustein, A. R., & Johnson, P. T. J. 2003. The complexity of deformed amphibians. *Frontiers in Ecology and the Environment* 1, 87–94.

Blumberg, M. S. 2002. *Body heat: Temperature and life on Earth.* Cambridge, MA: Harvard University Press.

———. 2005. *Basic instinct: The genesis of behavior.* New York: Thunder's Mouth Press.

Blumberg, M. S., & Wasserman, E. A. 1995. Animal mind and the argument from design. *American Psychologist* 50, 133–144.

Blumberg-Feldman, H., & Eilam, D. 1995. Postnatal development of synchronous stepping in the gerbil (*Gerbillus dasyurus*). *Journal of Experimental Biology* 198, 363–372.

Bogaert, A. F. 2006. Biological versus nonbiological older brothers and men's sexual orientation. *Proceedings of the National Academy of Sciences* 103, 10771–10774.

Bramble, D. M., & Lieberman, D. E. 2004. Endurance running and the evolution of *Homo. Nature* 432, 345–352.

Brenner, C. D. 1992. Electric limbs for infants and pre-school children. *Journal of Prosthetics and Orthotics* 4, 184–190.

Bronchti, G., Heil, P., Sadka, R., Hess, A., Scheich, H., & Wollberg, Z. 2002. Auditory activation of "visual" cortical areas in the blind mole rat (*Spalax ehrenbergi*). *European Journal of Neuroscience* 16, 311–329.

Bryant, S. V., Endo, T., & Gardiner, D. M., 2002. Vertebrate limb regeneration and the origin of limb stem cells. *International Journal of Developmental Biology* 46, 887–896.

Burke, A. C. 1989. Development of the turtle carapace: Implications for the evolution of a novel Bauplan. *Journal of Morphology* 199, 363–378.

Burke, A. C., & Nowicki, J. L. 2001. Hox genes and axial specification in vertebrates. *American Zoologist* 41, 687–697.

Burton, A. W. 1999. Hrdlicka (1931) revisited: Children who run on all fours. *Research Quarterly for Exercise and Sport* 70, 84–90.

Bush, G. L. 1982. Goldschmidt's follies. *Paleobiology* 8, 463–469.

Capel, B., & Coveney, D. 2004. Frank Lillie's freemartin: illuminating the pathway to 21st century reproductive endocrinology. *Journal of Experimental Zoology* 301, 853–856.

Carles, D., Weichhold, W., Alberti, E. M., Leger, F., Pigeau, F., & Horovitz, J. 1995. Diprosopia revisited in light of the recognized role of neural crest cells in facial development. *Journal of Craniofacial Genetics and Developmental Biology* 15, 90–97.

Carmena, J. M., Lebedev, M. A., Crist, R. E., O'Doherty, J. E., Santucci, D. M., Dimitrov, D. F., Patil, P. G., Henriquez, C. S., & Nicolelis, M. A. 2003. Learning to control a brain-machine interface for reaching and grasping by primates. *PLoS Biology* 1,193–208.

Caro, T. M., & Bateson, P. 1986. Organization and ontogeny of alternative tactics. *Animal Behaviour* 34, 1483–1499.

Carroll, L. 1960. *The annotated Alice*. New York: Meridian.

Carroll, S. B. 2005. *Endless forms most beautiful: The new science of Evo Devo*. New York: W. W. Norton.

Catania, K. C., & Henry, E. C. 2006. Touching on somatosensory specializations in mammals. *Current Opinion in Neurobiology* 16, 467–473.

Catania, K. C., & Kaas, J. H. 1997. Somatosensory fovea in the star-nosed mole: behavioral use of the star in relation to innervation patterns and cortical representation. *Journal of Comparative Neurology* 387, 215–233.

———. 1997. The mole nose instructs the brain. *Somatosensory & Motor Research* 14, 56–58.

Catania, K. C., Northcutt, R. G., & Kaas. J. H., 1999. The development of a biological novelty: a different way to make appendages as revealed in the snout of the star-nosed mole *Condylura cristata*. *Journal of Experimental Biology* 202, 2719–2726.

Catania, K. C., & Remple, F. E. 2005. Asymptotic prey profitability drives star-nosed moles to the foraging speed limit. *Nature* 433, 519–522.

Cattet, M. 1988. Abnormal sexual differentiation in black bears (*Ursus americanus*) and brown bears (*Ursus arctos*). *Journal of Mammalogy* 69, 849–852.

Cebra-Thomas, J., Tan, F., Sistla, S., Estes, E., Bender, G., Kim, C., Riccio, P., & Gilbert, S. F. 2005. How the turtle forms its shell: a paracrine hypothesis of carapace formation. *Journal of Experimental Biology* 304, 558–569.

Chiang, C., Litingtung, Y., Lee, E., Young, K. E., Corden, J. L., Westphal, H., & Beachy, P. A. 1996. Cyclopia and defective axial patterning in mice lacking *Sonic hedgehog* gene function. *Nature* 383, 407–413.

Clark, S. A., Allard, T., Jenkins, W. M., & Merzenich, M. M. 1988. Receptive fields in the body-surface map in adult cortex defined by temporally correlated inputs. *Nature* 332, 444–445.

Cohen, L. G., Celnik, P., Pascual-Leone, A., Corwell, B., Falz, L., Dambrosia, J., et al. 1997. Functional relevance of cross-modal plasticity in blind humans. *Nature* 389, 180–3.

Cohn, M. J., Lovejoy, C. O., L. Wolpert., & M. I. Coates. 2002. Branching, segmentation and the metapterygial axis: pattern versus process in the vertebrate limb. *BioEssays* 24, 460–465.

Cohn, M. J., & C. Tickle. 1999. Developmental basis of limblessness and axial patterning in snakes. *Nature* 399, 474–479.

Colapinto, J. 2004. Gender gap: What were the real reasons behind David Reimer's suicide? *Slate*, June 3.

Colton, H. S. 1929. How bipedal habit affects the bones of the hind legs of the albino rat. *Journal of Experimental Biology* 53, 1–11.

Constantine-Paton, M., & Law, M. I. 1978. Eye-specific termination bands in tecta of three-eyed frogs. *Science* 202, 639–641.

Conte, F. A., Grumbach, M. M., Ito, Y., Fisher, C. R., & Simpson, E. R. 1994. A syndrome of female pseudohermaphrodism, hyper-gonadotropic hypogonadism, and multicystic ovaries associated with missense mutations in the gene encoding aromatase (P450arom). *Journal of Clinical Endocrinology and Metabolism* 78, 1287–1292.

Cooper, M. K. 2004. Regenerative medicine: stem cells and the science of monstrosity. *Journal of Medical Ethics* 30, 12–22.

Cooper, M. K., Porter, J. A., Young, K. E., & Beachy P. A. 1998. Teratogen-mediated inhibition of target tissue response to *Shh* signaling. *Science* 280, 1603–1607.

Cordero, D., Marcucio, R., Hu, D., Gaffield, W., Tapadia, M., & Helms, J. A. 2004. Temporal perterbations in sonic hedgehog signaling elicit the spectrum of holoprosencephaly phenotypes. *Journal of Clinical Investigation* 114, 485–494.

Coventry, S., Kapur, R. P., & Siebert, J. R. 1998. Cyclopamine-induced holoprosencephaly and associated craniofacial malformations in the golden hamster: anatomic and molecular events. *Pediatric and Developmental Pathology* 1, 29–41.

Crews, D. 2003. Sex determination: where environment and genetics meet. *Evolution & Development* 5, 50–55.

———. 2005. Evolution of neuroendocrine mechanisms that regulate sexual behavior. *Trends in Endocrinology and Metabolism* 16, 354–361.

Crews, D., Bergeron, J. M., & McLachlan, J. A., 1995. The role of estrogen in turtle sex determination and the effect of PCBs. *Environmental Health Perspectives* 103 Suppl 7, 73–77.

Cunha, G. R., Place, N. J., Baskin, L., Conley, A., Weldele, M., Cunha, T. J., Wang, Y. Z., et al. 2005. The ontogeny of the urogenital system of the spotted hyena (*Crocuta crocuta Erxleben*). *Biology of Reproduction* 73, 554–564.

Dareste, C. 1891. *Recherches sur la production artificielle des monstruosités ou, essais de tératogenie expérimentale.* Paris: Reinwald.

Darwin, C. 1859/1983. *On the origin of species.* Harmondsworth: Penguin Books Ltd.

———. 1868/1998. *The variation of animals and plants under domestication.* Baltimore: Johns Hopkins University Press.

———. 1871. *The descent of man, and selection in relation to sex.* London: John Murray.

de Beer, G. R. 1940. *Embryos and ancestors.* Oxford: Clarendon Press.

de Guise, S., Lagace, A., & Beland, P., 1994. True hermaphroditism in a St. Lawrence beluga whale (*Delphinapterus leucas*). *Journal of Wildlife Diseases* 30, 287–90.

DeBrul, E. L., & Laskin, D. M. 1961. Preadaptive potentialities of the mammalian skull: An experiment in growth and form. *American Journal of Anatomy* 109, 117–132.

Diamond, M., & Sigmundson, H. K. 1997. Sex reassignment at birth: Long-term review and clinical implications. *Archives of Pediatrics and Adolescent Medicine* 15, 298–304.

Dove, W. F. 1936. Artificial production of the fabulous unicorn. *The Scientific Monthly* 2, 431–436.

Dreger, A. D. 1998. "Ambiguous sex" — or ambivalent medicine? Hastings Center Report 28, 24–35.

———. 2000. *Hermaphrodites and the medical invention of sex.* Cambridge: Harvard University Press.

Eilam, D. 1997. Postnatal development of body architecture and gait in several rodent species. *Journal of Experimental Biology* 200, 1339–1350.

Eilam, D., & Shefer, G.. 1997. The developmental order of bipedal locomotion in the jerboa (*Jaculus orientalis*): pivoting, creeping, quadrupedalism, and bipedalism. *Developmental Psychobiology* 31, 137–142.

Elbert, T., Pantev, C., Wienbruch, C., Rockstroh, B., & Taub, E. 1995. Increased cortical representation of the fingers of the left hand in string players. *Science* 270, 305–307.

Elbert, T., Sterr, A., Flor, H., Rockstroh, B., Knecht, S., Pantev, C., Wienbruch, C., & Taub, E. 1997. Input-increase and input-decrease types of cortical reorganization after upper extremity amputation in humans. *Experimental Brain Research* 117, 161–4.

Emlen, D. J. 2000. Integrating development with evolution: A case study with beetle horns. *BioScience* 50, 403–418.

———. 2001. Costs and the diversification of exaggerated animal structures. *Science* 291, 1534–1536.

Emlen, D. J., Marangelo, J., Ball, B., & Cunningham, C. W. 2005. Diversity in the weapons of sexual selection: horn evolution in the beetle genus *Onthophagus* (Coleoptera: Scarabaeidae). *Evolution* 59, 1060–1084.

Emlen, D. J., Szafran, Q., Corley, L. S., & Dworkin I. 2006. Insulin signaling and limb-patterning: candidate pathways for the origin and evolutionary diversification of beetle 'horns.' *Heredity* 97, 179–191.

Eugenides, J. 2002. *Middlesex*. New York: Picador.

Fagot, B. I., & Leinbach, M. D. 1989. The young child's gender schema: Environmental input, internal organization. *Child Development* 60, 663–672.

Fahrenthold, D. A. 2006. Male bass across region found to be bearing eggs. Washington Post, September 6, A01.

Fausto-Sterling, A. 1993. The five sexes. *The Sciences*, 20–25.

————. 2000. *Sexing the body: Gender politics and the construction of sexuality.* New York: Basic Books.

Fischer, D. H. 1970. *Historians' fallacies.* New York: Harper Torchbooks.

Fisher, R. A. 1930. *The genetical theory of natural selection.* Oxford: Clarendon Press.

Flor, H., Elbert, T., Muhlnickel, W., Pantev, C., Wienbruch, C., & Taub, E. 1998. Cortical reorganization and phantom phenomena in congenital and traumatic upper-extremity amputees. *Experimental Brain Research* 119, 205–212.

Frazzetta, T. H. 1970. From hopeful monsters to Bolyerine snakes? *The American Naturalist* 104, 55–72.

————. 1975. *Complex adaptations in evolving populations.* Sunderland, MA: Sinauer Associates, Inc.

Gaeth, A. P., Short, R. V., & Renfree, M. B. 1999. The developing renal, reproductive, and respiratory systems of the African elephant suggest an aquatic ancestry. *Proceedings of the National Academy of Sciences* 96, 5555–5558.

Galis, F., Wagner, G. P., & Jockusch, E. L. 2003. Why is limb regeneration possible in amphibians but not in reptiles, birds, and mammals? *Evolution & Development* 5, 208–220.

Garstang, W. 1922. The theory of recapitulation: A critical re-statement of the Biogenetic Law. *Journal of the Linnean Society, Zoology* 35, 81–101.

Gerhart, J. C., Black, S. D., Scharf, S., Gimlich, R., Vincent, J.-P., Danilchik, M., et al. 1986. Amphibian early development. *BioScience* 36, 541–549.

Gilbert, S. F. 2003. The morphogenesis of evolutionary developmental biology. *International Journal of Developmental Biology* 47, 467–477.

Gilbert, S. F., Loredo, G. A., Brukman, A., & Burke, A. C. 2001. Morphogenesis of the turtle shell: the development of a novel structure in tetrapod evolution. *Evolution & Development* 3, 47–58.

Gilbert, S. F., & Z. Zevit. 2001. Congenital human baculum deficiency: The generative bone of Genesis 2, 21–23. *American Journal of Medical Genetics* 101, 284–285.

Glickman, S. E. 1995. The spotted hyena from Aristotle to the Lion King: Reputation is everything. Social Research 62. Retrieved July 10, 2000, from http://findarticles.com/p/articles/mi_m2267/is_n3_v62/ai_17909878?tag=artBody;col1.

Glickman, S. E., Cunha, G. R., Drea, C. M., Conley, A. J., & Place, N. J. 2006. Mammalian sexual differentiation: lessons from the spotted hyena. *Trends in Endocrinology and Metabolism* 17, 349–356.

Glickman, S. E., Short, R. V., & Renfree, M. B. 2005. Sexual differentiation in three unconventional mammals: spotted hyenas, elephants and tammar wallabies. *Hormones and Behavior* 48, 403–417.

Godwin, J., Luckenbach, J. A., & Borski, R. J. 2003. Ecology meets endocrinology: environmental sex determination in fishes. *Evolution & Development* 5, 40–49.

Goldschmidt, R. 1933. Some aspects of evolution. *Science* 78, 539–547.

———. 1940. *The material basis of evolution.* New Haven: Yale University Press.

Gottlieb, G. 1992. *Individual development and evolution.* New York: Oxford University Press.

———. 1997. *Synthesizing nature-nurture: Prenatal roots of instinctive behavior.* Mahway: Lawrence Erlbaum Associates.

Gould, G. M., & Pyle, W. L. 1896. *Anomalies and curiosities of medicine.* New York: W. B. Saunders.

Gould, S. J. 1977. *Ontogeny and phylogeny.* Cambridge: The Belknap Press of Harvard University Press.

———. 1982. The uses of heresy: An introduction to Richard Goldschmidt's *The material basis of evolution.* In *The material basis of evolution* (pp. xiii–xlii) New Haven: Yale University Press.

————. 1995. The monster's human nature. In *Dinosaur in a haystack*. (pp. 53–62) New York: Harmony Books.

Gould, S. J., & Lewontin, R. C. 1979. The spandrels of San Marco and the Panglossian paradigm: a critique of the adaptationist programme. *Proceedings of the Royal Society of London* B205, 581–598.

Graham, J. M., Jr., Donahue, K. C., & Hall, J. G. 1993. Human anomalies and cultural practices. In *Human malformations and related anomalies* (pp. 169–181), R. E. Stevenson, J. G. Hall & R. M. Goodman (Eds.) New York: Oxford University Press.

Gross, M. R. 1991. Salmon breeding behavior and life history evolution in changing environments. *Ecology* 72, 1180–1186.

————. 1996. Alternative reproductive strategies and tactics: diversity within sexes. *Trends in Ecology and Evolution* 11, 92–98.

Gross, M. R., & Charnov, E. L. 1980. Alternative male life histories in bluegill sunfish. *Proceedings of the National Academy of Sciences* 77, 6937–6940.

Heil, P., Bronchti, G., Wollberg, Z., & Scheich, H. 1991. Invasion of visual cortex by the auditory system in the naturally blind mole rat. *Neuroreport* 2, 735–738.

Helms, J. A., Brugmann, S., & Cordero, D. R. (2008). SHH and other genes and the holoprosencephaly malformation sequence. In *Inborn errors of development*, Second edition (pp. 291–300), C. J. Epstein, R. P. Erickson & A. Wynshaw-Boris (Eds.) New York: Oxford University Press.

Helms, J. A., Cordero, D., & Tapadia, M. D. 2005. New insights into craniofacial morphogenesis. *Development* 132, 851–61.

Hirasaki, E., Ogihara, N., Hamada, Y., Kumakura, H., & Nakatsukasa, M. 2004. Do highly trained monkeys walk like humans? A kinematic study of bipedal locomotion in bipedally trained Japanese macaques. *Journal of Human Evolution* 46, 739–750.

Hölldobler, B., & Wilson, E. O. 1990. *The ants*. Cambridge, MA: Harvard University Press.

Houk, C. P., Hughes, I. A., Ahmed, S. F., & Lee, P. A. 2006. Summary of consensus statement on intersex disorders and their management. *Pediatrics* 118, 753–757.

Hoyt, D. F., & Taylor, C. R. 1981. Gait and the energetics of locomotion in horses. *Nature* 292, 239–240.

Hu, D., & Helms, J. A. 1999. The role of sonic hedgehog in normal and abnormal craniofacial morphogenesis. *Development* 126, 4873–4884.

Humphrey, N., Skoyles, J. R., & Keynes, R. 2005. Human hand-walkers: Five siblings who never stood up. In Centre for Philosophy of Natural and Social Science Discussion Paper.

Hunt, J., & Simmons, L. W. 2000. Maternal and paternal effects on off-spring phenotype in the dung beetle *Onthophagus taurus*. *Evolution* 54, 936–941.

Hunter, J. 1779. Account of the free martin. *Philosophical Transactions of the Royal Society of London* 69, 279–293.

Huxley, T. H. 1863. *Evidence as to man's place in nature*. London: Williams & Norgate.

Jablonka, E., & Lamb, M. J. 2002. The changing concept of epigenetics. *Annals of the New York Academy of Sciences* 981, 82–96.

———. 2005. *Evolution in four dimensions: Genetic, epigenetic, behavioral, and symbolic variation in the history of life*. Cambridge: MIT Press.

Jacob, F. 1977. Evolution and tinkering. *Science* 196, 1161–1166.

Johnson, P. T. J., Lunde, K. B., Ritchie, E. G., & Launer, A. E. 1999. The effect of trematode infection on amphibian limb development and survivorship. *Science* 284, 802–804.

Johnson, P. T. J., Preu, E. R., Sutherland, D. R., Romansic, J. M., Han, B., & Blaustein, A. R. 2006. Adding infection to injury: synergistic effects of predation and parasitism on amphibian malformations. *Ecology* 87, 2227–2235.

Kageura, H. 1997. Activation of dorsal development by contact between the cortical dorsal determinant and the equatorial core cytoplasm in eggs of *Xenopus laevis*. *Development* 124, 1543–1551.

Kahn, D. M., & Krubitzer, L. 2002. Massive cross-modal cortical plasticity and the emergence of a new cortical area in developmentally blind mammals. *Proceedings of the National Academy of Sciences* 99, 11429–11434.

Karlen, S. J., Kahn, D. M., & Krubitzer, L. 2006. Early blindness results in abnormal corticocortical and thalamocortical connections. *Neuroscience* 142, 843–858.

Kaufman, M. H. 2004. The embryology of conjoined twins. *Child's Nervous System* 20, 508–525.

Keeler, R. F. 1970. Teratogenic compounds of *Veratrum californicum* (Durand) X. Cyclopia in rabbits produced by cyclopamine. *Teratology* 3, 175–180.

Keeler, R. F., & Binns, W. 1968. Teratogenic compounds of *Veratrum californicum* (Durand). V. Comparison of cyclopian effects of steroidal alkaloids from the plant and structurally related compounds from other sources. *Teratology* 1, 5–10.

Keller, E. F. 2000. *The century of the gene.* Cambridge, MA: Harvard University Press.

Kiesecker, J. M. 2002. Synergism between trematode infection and pesticide exposure: A link to amphibian limb deformities in nature? *Proceedings of the National Academy of Sciences* 99, 9900–9904.

Killackey, H. P., Rhoades, R. W., & Bennett-Clarke, C. A.. 1995. The formation of a cortical somatotopic map. *Trends in Neurosciences* 18, 402–407.

Kirschner, M. W., & Gerhart, J. C. 2005. *The plausibility of life: Resolving Darwin's dilemma.* New Haven: Yale University Press.

Krubitzer, L, & Kaas, J. 2005. The evolution of the neocortex in mammals: how is phenotypic diversity generated? *Current Opinion in Neurobiology* 15, 444–453.

Kuiper, H., Bunck, C., Gunzel-Apel, A. R., Drogemuller, C., Hewicker-Trautwein, M., & Distl, O. 2005. SRY-negative XX sex reversal in a Jack Russell Terrier: a case report. *Veterinary Journal* 169, 116–117.

Le Guyader, H. 1998. *Etienne Geoffroy Saint-Hilaire, 1772–1844: A visionary naturalist.* Chicago: University of Chicago Press.

Lebedev, M. A., & Nicolelis, M. A. 2006. Brain-machine interfaces: past, present and future. *Trends in Neurosciences* 29, 536–546.

Lehrman, D. S. 1953. A critique of Konrad Lorenz's theory of instinctive behavior. *The Quarterly Review of Biology* 4, 337–363.

Leroi, A. M. 2003. *Mutants: On genetic variety and the human body.* New York: Penguin.

Lesbre, F.-X. 1927. *Traité do Tératologie de l'Homme et des animaux domestiques.* Paris: Vigot Fréres.

Lewontin, R. 1992. Genotype and phenotype. In *Keywords in evolutionary biology* (pp. 137–144), E. F. Keller & E. A. Lloyd (Eds.) Cambridge: Harvard University Press.

Li, H., Tierney, C., Wen, L., Wu, J. Y., & Rao Y. 1997. A single morpho-genetic field gives rise to two retina primordia under the influence of the prechordal plate. *Development* 124, 603–615.

Lillie, F. R. 1916. The theory of the free-martin. *Science* 43, 611–613.

Lotze, M., Grodd, W., Birbaumer, N., Erb, M., Huse, E., & Flor, H. 1999. Does use of a myoelectric prosthesis prevent cortical reorganization and phantom limb pain? *Nature Neuroscience* 2, 501–502.

Love, A. C. 2003. Evolutionary morphology, innovation, and the synthesis of evolutionary and developmental biology. *Biology and Philosophy* 18, 309–345.

Marcucio, R. S., Cordero, D. R,. Hu, D., & Helms, J. A. 2005. Molecular interactions coordinating the development of the forebrain and face. *Developmental Biology* 284, 48–61.

Maynard Smith, J., Burian, R., Kauffman, S., Alberch, P., Campbell, J., Goodwin, B., et al. 1985. Developmental constraints and evolution. *The Quarterly Review of Biology* 60, 265–287.

Mayor, A. 2000. *The first fossil hunters: Paleontology in Greek and Roman times.* Princeton: Princeton University Press.

Mayr, E. 1976. Behavior programs and evolutionary strategies. In *Evolution and the diversity of life* (pp. 694–711) Cambridge, MA: Harvard University Press.

McGann, C. J., Odelberg, S. J., & Keating, M. T. 2001. Mammalian myotube dedifferentiation induced by newt regeneration extract. *Proceedings of the National Academy of Sciences* 98, 13699–13704.

McHugh, P. R. 1995. Witches, multiple personalities, and other psychiatric artifacts. *Nature Medicine* 1, 110–114.

———. 2004. Surgical sex. First Things: The Journal of Religion, Culture, and Public Life November. Retrieved June 5, 2007, from http://www.firstthings.com/article.php3?id_article=398.

McNamara, K. J. 1997. *Shapes of life: The evolution of growth and development.* Baltimore: Johns Hopkins University Press.

Merzenich, M. M., Nelson, R. J., Stryker, M. P., Cynader, M. S., Schoppmann, A., & Zook, J. M. 1984. Somatosensory cortical map changes following digit amputation in adult monkeys. *Journal of Comparative Neurology* 224, 591–605.

Metcalfe, N. B., Huntingford, F. A., Graham, W. D., & Thorpe, J. E. 1989. Early social status and the development of life-history strategies in Atlantic salmon. *Proceedings of the Royal Society of London B* 236, 7–19.

Moczek, A. P., Cruickshank, T. E., & Shelby, A. 2006. When ontogeny reveals what phylogeny hides: Gain and loss of horns during development and evolution of horned beetles. *Evolution* 60, 2329–2341.

Moczek, A. P., & Emlen, D. J. 2000. Male horn dimorphism in the scarab beetle, *Onthophagus taurus*: do alternative reproductive tactics favour alternative phenotypes? *Animal Behaviour* 59, 459–466.

Modi, W. S., & Crews, D. 2005. Sex chromosomes and sex determination in reptiles. *Current Opinion in Genetics & Development* 15, 660–5.

Moore, D. S. 2001. *The dependent gene: The fallacy of "nature vs. nurture."* New York: W. H. Freeman & Company.

Moore, M. C. 1991. Application of organization-activation theory to alternative male reproductive strategies: A review. *Hormones and Behavior* 25, 154–179.

Muir, D. C., Backus, S., Derocher, A. E., Dietz, R., Evans, T. J., Gabrielsen, G. W., et al. 2006. Brominated flame retardants in polar bears (*Ursus maritimus*) from Alaska, the Canadian Arctic, East Greenland, and Svalbard. *Environmental Science & Technology* 40, 449–455.

Munday, P. L., Buston, P. M., & Warner, R. R. 2006. Diversity and flexibility of sex-change strategies in animals. *Trends in Ecology & Evolution* 21, 89–95.

Munday, P. L., Wilson White, J., & Warner, R. R.. 2006. A social basis for the development of primary males in a sex-changing fish. *Proceedings of the Royal Society of London B* 273, 2845–2851.

Neff, B. D., Fu, P., & Gross, M. R. 2003. Sperm investment and alternative mating tactics in bluefull sunfish (*Lepomis macrochirus*). *Behavioral Ecology* 14, 634–641.

Neiman, M. 2004. Physiological dependence on copulation in parthe-
nogenetic females can reduce the cost of sex. *Animal Behaviour* 67,
811–822.

Newman, S. A., & Müller, G. B. 2000. Epigenetic mechanisms of character origination. *Journal of Experimental Biology* 288, 304–317.

————. 2005. Origination and innovation in the vertebrate limb skeleton: An epigenetic perspective. *Journal of Experimental Biology* 304B,
593–609.

Nijhout, H. F. 1990. Metaphors and the role of genes in development.
BioEssays 12, 441–446.

O'Connor, E. 2000. *Raw material: Producing pathology in Victorian culture.*
Durham: Duke University Press.

Oppenheimer, J. M. 1968. Some historical relationships between teratology and experimental embryology. *Bulletin of the History of Medicine*
42, 145–159.

Oster, G. F., Shubin, N., Murray, J. D., & Alberch, P. 1988. Evolution
and morphogenetic rules: The shape of the vertebrate limb in
ontogeny and phylogeny. *Evolution* 42, 862–884.

Owusu-Fimpong, M., & Hargreaves, J. A. 2000. Incidence of conjoined
twins in tilapia after thermal shock induction of polyploidy.
Aquaculture Research 31, 421–426.

Oyama, S., Griffiths, P. E., & Gray, R. D. (Eds.). 2001. *Cycles of contingency:
Developmental systems and evolution.* Cambridge: MIT Press.

Papanicolaou, G. N., & Stockard, C. R. 1917. The existence of a typical
oestrous cycle in the guinea-pig—with a study of its histolog-
ical and physiological changes. *American Journal of Anatomy* 22,
225–283.

Papanicolaou, G. N., & Traut, H. F. 1941. The diagnostic value of vaginal smears in carcinoma of the uterus. *American Journal of Obstetrics
and Gynecology* 42, 193–206.

Paré, A. 1573/1982. *On monsters and marvels.* Chicago: University of Chicago Press.

Pauly, P. J. 1996. How did the effects of alcohol on reproduction become scientifically uninteresting? *Journal of the History of Biology* 29, 1–28.

Pearl, R. 1916. On the effect of continued administration of certain poisons to the domestic fowl, with special reference to the progeny. *Proceedings of the American Philosophical Society* 55, 243–258.

————. 1927. The biology of superiority. *The American Mercury*, November, 257–266.

Petersen, C. W. 1995. Reproductive behavior, egg trading, and correlates of male mating success in the simultaneous hermaphrodite, *Serraus tabacarius. Environmental Biology of Fishes* 43, 351–361.

Pfaff, D. W. 1997. Hormones, genes, and behavior. *Proceedings of the National Academy of Sciences* 94, 14213–14216.

Phelan, M. C. 1993. Twins. In *Human malformations and related anomalies* (pp. 1047–1079), R. E. Stevenson, J. G. Hall, & R. M. Goodman (Eds.) New York: Oxford University Press.

Pietsch, T. W. 2005. Dimorphism, parasitism, and sex revisited: Modes of reproduction among deep-sea ceratioid anglerfishes (Teleostei: Lophiiformes). *Ichthyological Research* 52, 207–236.

Pigliucci, M. 2001. *Phenotypic plasticity: Beyond nature and nurture.* Baltimore: Johns Hopkins University Press.

Place, N. J., & Glickman, S. E. 2004. Masculinization of female mammals: lessons from nature. *Advances in Experimental Medicine and Biology* 545, 243–253.

Provine, W. B. 1971. *The origins of theoretical population genetics.* Chicago: University of Chicago Press.

Puts, D. A., Jordan, C. L., & Breedlove, S. M. 2006. O brother, where art thou? The fraternal birth-order effect on male sexual orientation. *Proceedings of the National Academy of Sciences* 103, 10531–10532.

Quinn, A. E., Georges, A., Sarre, S. D., Guarino, F., Ezaz, T., & Graves, J. A. 2007. Temperature sex reversal implies sex gene dosage in a reptile. *Science* 316, 411.

Rachootin, S. P., & Thomson, K. S. 1981. Epigenetics, paleontology, and evolution. In *Evolution today: Proceedings of the second international congress of systematic and evolutionary biology* (pp. 181–193), G. G. E. Scudder & J. E. Reveal (Eds.) Pittsburgh: Hunt Institute for Botanical Documentation.

Reiner, W. G., & Gearhart, J. P. 2004. Discordant sexual identity in some genetic males with cloacal exstrophy assigned to female sex at birth. *New England Journal of Medicine* 350, 333–341.

Richards, E. 1994. A political anatomy of monsters, hopeful and otherwise. *Isis* 85, 377–411.

Richards, R. J. 1992. The meaning of evolution. Chicago: University of Chicago Press.

Richardson, M. K. 1999. Vertebrate evolution: the developmental origins of adult variation. *BioEssays* 21, 604–613.

Richardson, M. K., Hanken, J., Gooneratne, M. L., Pieau, C., Raynaud, A., Selwood, L. 1997. There is no highly conserved embryonic stage in the vertebrates: implications for current theories of evolution and development. *Anatomy and Embryology* 196, 91–106.

Rieppel, O. 2001. Turtles as hopeful monsters. *BioEssays* 23, 987–991.

Robert, J. S. 2002. How developmental is evolutionary developmental biology? *Biology and Philosophy* 17, 591–611.

Robinson, S. R. 2005. Conjugate limb coordination after experience with an interlimb yoke: evidence for motor learning in the rat fetus. *Developmental Psychobiology* 47, 328–344.

Robinson, S. R., & Kleven, G. A. 2005. Learning to move before birth. In *Advances in infancy research: Prenatal development of postnatal functions* (pp. 131–175), B. Hopkins & S. P. Johnson (Eds.) Westport, CT: Praeger Publishers.

Roder, B., Teder-Salejarvi, W., Sterr, A., Rosler, F., Hillyard, S. A. 1999. Improved auditory spatial tuning in blind humans. *Nature* 400, 162–166.

Rodgers, E. W., Earley, R. L., & Grober, M. S. 2007. Social status determines sexual phenotype in the bi-directional sex changing bluebanded goby *Lythrypnus dalli*. *Journal of Fish Biology* 70, 1660–1668.

Roessler, E., & Muenke, M. 1998. Holoprosencephaly: A paradigm for the complex genetics of brain development. *Journal of Inherited Metabolic Disease* 21, 481–497.

Rosenberg, D. 2007. Rethinking gender. *Newsweek*, May 21, 50–57.

Sarre, S. D., Georges, A., & Quinn, A. 2004. The ends of a continuum: genetic and temperature-dependent sex determination in reptiles. *Bioessays* 26, 639–45.

Scanlon, P. F., Urbson, D. F., & Sullivan, J. A. 1975. A male pseudohermaphrodite white-tailed deer resembling an antlered doe. *Journal of Wildlife Diseases* 11, 237–40.

Schneider, R. A. 2005. Developmental mechanisms facilitating the evolution of bills and quills. *Journal of Anatomy* 207, 563–573.

Semsar, K., & Godwin, J. 2003. Social influences on the arginine vasotocin system are independent of gonads in a sex-changing fish. *Journal of Neuroscience* 23, 4386–4393.

Shapiro, M. D., Marks, M. E., Peichel, C. L., Blackman, B. K., Nereng, K. S., Jonsson, B. 2004. Genetic and developmental basis of evolutionary pelvic reduction in threespine sticklebacks. *Nature* 428, 717–723.

Shozu, M., Akasofu, K., Harada, T., & Kubota, Y. 1991. A new cause of female pseudohermaphroditism: placental aromatase deficiency. *Journal of Clinical Endocrinology and Metabolism* 72, 560–566.

Shubin, N. H., & Alberch, P. 1986. A morphogenetic approach to the origin and basic organization of the tetrapod limb. In *Evolutionary biology* (pp. 319–387), M. K. Hecht, B. Wallance, & G. T. Prance (Eds.) New York: Plenum Press.

Slijper, E. J. 1942a. Biologic-anatomical investigations on the bipedal gait and upright posture in mammals, with specieal reference to a little goat, born without forelegs. I. Proc. *Koninklijke Nederlandse Akademie Van Wetenschappen* 5, 288–295.

————. 1942b. Biologic-anatomical investigations on the bipedal gait and upright posture in mammals, with specieal reference to a little goat, born without forelegs. II. Proc. *Koninklijke Nederlandse Akademie Van Wetenschappen* 5, 407–415.

Sonne, C., Leifsson, P. S., Dietz, R., Born, E. W., Letcher, R. J., Kirkegaard, M. 2005. Enlarged clitoris in wild polar bears (*Ursus maritimus*) can be misdiagnosed as pseudohermaphroditism. *Science of the Total Environment* 337, 45–58.

Spelke, E. S., & Kinzler, K. D. 2007. Core knowledge. *Developmental Science* 10, 89–96.

Spencer, J. P., Blumberg, M. S., McMurray, R., Robinson, S. R., Samuelson, L. K., & Tomblin, J. B. (in press). Short arms and talking eggs: Why we should no longer abide the nativist-empiricist debate. *Child Development Perspectives*.

Spencer, R. 2000. Theoretical and analytical embryology of conjoined twins: Part I: Embryogenesis. *Clinical Anatomy* 13, 36–53.

————. 2003. *Conjoined twins: Developmental malformations and clinical implications.* Baltimore: Johns Hopkins University Press.

Stamper, M. A., Norton, T., Spodnick, G., Marti, J., & Loomis. M. 1999. Hypospadias in a polar bear (*Ursus maritimus*). *Journal of Zoo and Wildlife Medicine* 30, 141–144.

Stevenson, R. E. 1993. Causes of human anomalies: An overview and historical perspective. In *Human malformations and related anomalies* (pp. 3–20), R. E. Stevenson, J. G. Hall, & R. M. Goodman (Eds.) New York: Oxford University Press.

Stockard, C. R. 1912. An experimental study of racial degeneration in mammals treated with alcohol. *Archives of Internal Medicine* 10, 369–398.

————. 1912. Is the control of the embryonic development a practical problem? *Proceedings of the American Philosophical Society* 51, 191–200.

————. 1921. Developmental rate and structural expression: An experimental study of twins, 'double monsters' and single deformities, and the interaction among embryonic organs during their origin and development. *American Journal of Anatomy* 28, 115–277.

————. 1941. *The genetic and endocrinic basis for differences in form and behavior.* Philadelphia: The Wistar Institute of Anatomy and Biology.

Stockard, C. R., & Papanicolaou, G. N. 1916. A further analysis of the hereditary transmission of degeneracy and deformities by the descendants of alcoholized mammals. *The American Naturalist* 50, 65–88.

Stopper, G. F., Hecker, L., Franssen, R. A., & Sessions, S. K. 2002. How trematodes cause limb deformities in amphibians. *Journal of Experimental Biology* 294, 252–63.

Taborsky, M. 1994. Sneakers, satellites, and helpers: Parasitic and cooperative behavior in fish reproduction. *Advances in the Study of Behavior* 23, 1–100.

ten Cate, C. 1994. Perceptual mechanisms in imprinting and song learning. In *Causal mechanisms of behavioural development* (pp. 116–146), J. A. Hogan & J. J. Bolhuis (Eds.) Cambridge: Cambridge University Press.

Thewissen, J. G., Cohn, M. J., Stevens, L. S., Bajpai, S., Heyning, J., & Horton, W. E. Jr. 2006. Developmental basis for hind-limb loss in dolphins and origin of the cetacean bodyplan. *Proceedings of the National Academy of Sciences* 103, 8414–8.

Thorpe, J. E., Metcalfe, N. B., & Huntingford, F. A. 1992. Behavioral influences on life-history variation in juvenile Atlantic salmon, *Salmo salar*. *Environmental Biology of Fishes* 33, 331–340.

Tickle, C. 2002. The early history of the polarizing region: from classical embryology to molecular biology. *International Journal of Developmental Biology* 46, 847–52.

Ulshafer, R. J., & Clavert, A. 1979. The use of avian double monsters in studies on induction of the nervous system. *Journal of Embryology and Experimental Morphology* 53, 237–243.

Vance, S. A. 1996. Morphological and behavioral sex reversal in mermithid-infected mayflies. *Proceedings of the Royal Society of London* B 263, 907–912.

Vandenbergh, J. G. 2003. Prenatal hormone exposure and sexual variation. *American Scientist* 91, 218–225.

Vaughan, L., Schofield, W., & Ennis, S. 2001. SRY-negative XX sex reversal in a pony: a case report. *Theriogenology* 55, 1051–1057.

Viala, D., Viala, G., & Fayein, N. 1986. Plasticity of locomotor organization in infant rabbits spinalized shortly after birth. In *Development and plasticity of the mammalian spinal cord* (pp. 301–310), M. E. Goldberger, A. Gorio, & M. Murray (Eds.) Padova: Liviana Press.

Waddington, C. H. 1957. *The strategy of the genes*. London: Allen & Unwin.

Wake, D. B. 1998. Pere Alberch (1954–98). *Nature* 393, 632.

Wasserman, E. A., & Blumberg, M. S. 2006. Designing minds. *APS Observer* 19, 25–26.

Weil, E. 2006. What if it's (sort of) a boy and (sort of) a girl? *New York Times Magazine*, September 24.

Wells, H. G. 1896/2004. *The island of Dr. Moreau.* New York: Barnes & Noble Books.

West, M. J., King, A. P., & Arberg, A. A. 1988. The inheritance of niches: The role of ecological legacies in ontogeny. In *Handbook of behavioral neurobiology* (pp. 301–310), E. M. Blass. (Ed.) New York: Plenum Press.

West-Eberhard, M. J. 2003. *Developmental plasticity and evolution.* Oxford: Oxford University Press.

———. 2005. Developmental plasticity and the origin of species differences. *Proceedings of the National Academy of Sciences* 102 Suppl 1, 6543–6549.

Wewerka, S., & Miller, K. 1996. Together forever. *Life* April, 44–54.

Wiig, O., Derocher, A. E., Cronin, M. M., & Skaare, J. U. 1998. Female pseudohermaphrodite polar bears at Svalbard. *Journal of Wildlife Diseases* 34, 792–796.

Wilder, H. H. 1904. Duplicate twins and double monsters. *American Journal of Anatomy* 3, 387–3472.

———. 1908. The morphology of cosmobia; speculations concerning the significance of certain types of monsters. *American Journal of Anatomy* 8, 355–440.

Wilson, P. K. 2003. Bad habits and bad genes: Early 20th-century eugenic attempts to eliminate syphilis and assocated "defects" from the United States. *Canadian Bulletin of Medical History* 20, 11–41.

Woodhouse, C. R. 1998. Sexual function in boys born with exstrophy, myelomeningocele, and micropenis. *Urology* 52, 3–11.

Wu, J., Staffenberg, D. A., Mulliken, J. B., & Shanske, A. L. 2002. Diprosopus: a unique case and review of the literature. *Teratology* 66, 282–287.

Zardoya, R., & Meyer, A. 2001. The evolutionary position of turtles revised. *Naturwissenschaften* 88, 193–200.

AUTHOR INDEX

SUBJECT INDEX